Vegetable and Fruit Farming: The Complete Guide

Vegetable and Fruit Farming: The Complete Guide

Kenneth Lewis

STATES
ACADEMIC PRESS
www.statesacademicpress.com

States Academic Press,
109 South 5th Street,
Brooklyn, NY 11249, USA

Visit us on the World Wide Web at:
www.statesacademicpress.com

ISBN: 978-1-63989-555-7 (Hardback)

Cataloging-in-Publication Data

Vegetable and fruit farming : the complete guide / Kenneth Lewis.
    p. cm.
Includes bibliographical references and index.
ISBN 978-1-63989-555-7
1. Vegetable gardening. 2. Fruit-culture. 3. Vegetables. 4. Fruit.
5. Horticulture. 6. Gardening. I. Lewis, Kenneth.
SB320.9 .V44 2022
635--dc23

Table of Contents

Preface

The practice of growing crops for obtaining food and raw materials is called farming. The discipline of vegetable and fruit farming focuses on producing different fruits and vegetables for human consumption. There are various climatic factors which influence vegetable and food production such as temperature, moisture, and daylight. Various modern day techniques used for growing fruits and vegetables include organic farming, vertical farming, hydroponic farming and raised bed farming. Some of the important aspects of fruit cultivation are soil management, irrigation, pollination, thinning, pruning and spacing systems. This book is compiled in such a manner, that it will provide in-depth knowledge about the theory and practice of vegetable and fruit farming. Different approaches, evaluations and methodologies on vegetable and fruit farming have been included herein. It will serve as a valuable source of knowledge for those interested in this field.

A detailed account of the significant topics covered in this book is provided below:

Chapter 1- The growing of vegetables and fruits for human consumption is termed as vegetable and fruit farming. The vegetables can be broadly categorized into leafy vegetables, root vegetables, stem vegetables and fruit vegetables. This is an introductory chapter which will introduce briefly all the significant aspects of fruits and vegetable farming.

Chapter 2- There are numerous vegetables which are grown for human consumption such as capsicum, tomato, cucumber, eggplant and pumpkin. A few major fruit crops are guava, banana, apple and strawberry. This chapter discusses in detail these vegetable and fruit crops as well as their farming.

Chapter 3- An orchard is a plantation of trees or shrubs that is maintained for the production of food. Orchards comprise fruit-producing and nut-producing trees which are generally grown for commercial purposes. This chapter has been carefully written to provide an easy understanding of the varied facets of orchards.

Chapter 4- Preservation slows down the activity of disease-causing bacteria in food. There are various methods of preserving vegetables and fruits such as freezing, drying and canning. These methods of preservation of vegetables and fruits as well as their storage have been thoroughly discussed in this chapter.

Chapter 5- The vegetable and fruit crops can be affected by a large number of diseases such as black dot, beet vascular necrosis, alternaria solani, tobamovirus, didymella pinodes, ralstonia solanacearum and phytophthora capsici. All these vegetables and fruits diseases have been carefully analyzed in this chapter.

It gives me an immense pleasure to thank our entire team for their efforts. Finally in the end, I would like to thank my family and colleagues who have been a great source of inspiration and support.

Kenneth Lewis

Chapter 1

Vegetable and Fruit Farming: An Introduction

The growing of vegetables and fruits for human consumption is termed as vegetable and fruit farming. The vegetables can be broadly categorized into leafy vegetables, root vegetables, stem vegetables and fruit vegetables. This is an introductory chapter which will introduce briefly all the significant aspects of fruits and vegetable farming.

Vegetable

Vegetable, in the broadest sense is any kind of plant life or plant product, namely "vegetable matter"; in common, narrow usage, the term vegetable usually refers to the fresh edible portions of certain herbaceous plants—roots, stems, leaves, flowers, fruit, or seeds. These plant parts are either eaten fresh or prepared in a number of ways, usually as a savory, rather than sweet, dish.

Assorted fresh vegetables.

Virtually all of the more important vegetables were cultivated among the ancient civilizations of either the Old or the New World and have long been noted for their nutritional importance. Most fresh vegetables are low in calories and have a water content in excess of 70 percent, with only about 3.5 percent protein and less than 1 percent fat. Vegetables are good sources of minerals, especially calcium and iron, and vitamins, principally A and C. Nearly all vegetables are rich in dietary fibre and antioxidants.

Variety of potatoes (*Solanum tuberosum*).

Vegetables are usually classified on the basis of the part of the plant that is used for food. The root vegetables include beets, carrots, radishes, sweet potatoes, and turnips. Stem vegetables include asparagus and kohlrabi. Among the edible tubers, or underground stems, are potatoes. The leaf and leafstalk vegetables include brussels sprouts, cabbage, celery, lettuce, rhubarb, and spinach. Among the bulb vegetables are garlic, leeks, and onions. The head, or flower, vegetables include artichokes, broccoli, and cauliflower. The fruits commonly considered vegetables by virtue of their use include cucumbers, eggplant, okra, sweet corn, squash, peppers, and tomatoes. Seed vegetables are usually legumes, such as peas and beans.

Unshelled peas.

Modern vegetable farming ranges from small-scale production for local sale to vast commercial operations utilizing the latest advances in automation and technology. In addition, vegetables can be grown conventionally or using organic farming methods. Most vegetables are planted by seeding in the fields where they are to be grown, but occasionally they are germinated in a nursery or greenhouse and transplanted as seedlings to the field. During the growing season synthetic or organic herbicides, pesticides, and fungicides are commonly used to inhibit damage by weeds, insects, and diseases, respectively. Depending on the crop, harvesting operations are usually mechanized in well-developed countries, but the practice of harvesting by hand is still employed in some areas or is used in conjunction with machine operations. Another concern of the vegetable farmer is postharvest storage, which may require refrigerated facilities.

Vegetables may be washed, sorted, graded, cut, and packaged for sale as fresh products. Fresh vegetables are subject to quick aging and spoilage, but their storage life can be extended by such preservation processes as dehydration, canning, freezing, fermenting, or pickling.

Leafy Vegetables

Leafy vegetables are the most nutritious because it contains high amounts of carotene, Riboflavin, Folic acid, Calcium and Iron. They also contain fibre which helps to prevent constipation and reduce the rate of absorption of dietary cholesterol.

Leafy vegetables such as amaranth, drumstick leaves, spinach, Fenugreek leaves can be easily cultivated and are inexpensive. They can be easily grown at home. They are best to be consumed when they are fresh and can be stored in fridge to keep them fresh for a longer duration of time.

They can be used to prepare other food items, in the form of stuffing and mixing as small children mostly don't relish the idea of having leafy vegetables. The addition of green leafy vegetables in your diet gives out lots of health benefits. Leafy greens are full of vitamins, minerals, and disease-fighting phytochemicals and helps the growth and repair of body tissues. Vitamin E and Vitamin C in leafy vegetables combine to give beautiful skin and healthy hair growth.

Other Important Health Benefits of having Leafy Vegetables

- Green leafy vegetables are a good source of fibre, which helps to reduce weight as the consumption would make you feel full and lessens your hunger level. Fibrous food items also contribute to lower your cholesterol and blood pressure.

- The risk of developing osteoporosis also comes down since some leafy vegetables are rich sources calcium in absorbable form. It also enhances muscle function and manages blood-pressure.

- Beta-carotene is an important component for your eyes and has antioxidant to support you with lots of health benefits for your skin and body. This element is present in other fruits and vegetables such as collards, spinach, mangoes, carrot, passion fruit etc.

Some of the most commonly consumed green leafy vegetables are:

- Cauliflower: It is a highly nutritious vegetable and contains vitamins such as thiamine (vit B1), riboflavin (B2), niacin, pantothenic acid, pyridoxine and folic acid. It also has essence of omega-3 fatty acids and vitamin K. Cauliflower also serves as a good source of proteins, phosphorus, potassium, vitamin C and manganese both works as powerful antioxidants.

- Cabbage: Highly nutritious vegetable which is low in fat and calories. It is a storehouse of phyto chemicals and acts as a powerful antioxidant and helps body develop resistance against infectious agents and disease causing free radicals. It also contains an adequate amount of minerals like potassium, manganese, iron, and magnesium.

- Broccoli: If broccoli is cooked by steaming, it can provide you with cholesterol-lowering benefits. It is also fibrous to some extent. Broccoli is a good carb and is high in fiber, which aids in digestion, prevents constipation, maintains low blood sugar, and limits overeating by making you, felt fuller.

- Spinach: Green leafy spinach leaves contains greater concentrations of vitamin C. Consumption of spinach includes improvement in blood glucose control in diabetics, lowering the risk of cancer, blood-pressure and also contributes to proper bone health.

- Coriander leaves: It is basically used for garnishing purpose. Coriander leaves are used at the end of cooking to add a flavour to the curry. They are the best source for carrying out proper digestion. Also it has got high amount of iron supplements thus helps prevent anaemia.

Root Vegetables

Root vegetables are quite literally the fruits of the earth, the hearty bulbs that swell and thrive beneath the soil. Flashier plants and trees will suspend networks of roots below ground, catching nutrients from the soil and using this energy to grow up and out. Root vegetables do things a little differently. Unlike eye-catching floral blooms or fragrant herbs, where the leaves are the coveted parts of the plant, root vegetables prize—you guessed it—the hard-working roots.

The root vegetables themselves absorb nutrients from the soil, fortifying them as some of the heartiest, healthiest foods you can find in the fall. Plus, pulling up your root vegetables at harvest time to reveal full orange carrots or bright beets is as exciting as discovering buried treasure.

Types of Root Vegetables

Root vegetables come in all shapes, colors, and sizes. Some of the most common root vegetables are:

- Potatoes

- Carrots

- Beets

- Turnips

- Parsnips

- Onions and shallots

- Radishes

- Sweet potatoes

How to Grow Root Vegetables

Root vegetables are relatively easy to grow and, in the case of an abundant fall harvest, reap plentiful rewards. To grow root vegetables, you'll need a loose, ideally raised soil bed. The key to growing root vegetables is ensuring that the soil is loose enough for the plants to send down roots. They grow best in cool weather, so you'll want to plant your vegetables in mid to late summer for a fall harvest. Full sun exposure is ideal for growing root vegetables.

When growing root vegetables, it's important to be meticulous and intentional in their spacing. Root vegetables need space to grow, and since the seeds are so small, you may have to wait a few weeks after first sowing and go back and adjust the plants' spacing. Ideally, you want to space the vegetables 2 to 4 inches apart (potatoes, however, will need much more space).

Stem Vegetables

Stem vegetables are those that have shoots or stalks which can be consumed. Some of the most popular stem vegetables include asparagus, celery, fennel, etc. These vegetables can be used to make a variety of dishes and are usually served with pasta, sandwiches, soups, etc. It is important to remember that stem vegetables should not be over-cooked, or else they may lose their crunchiness. Along with being delicious, stem vegetables possess minerals, vitamins and antioxidants.

List of Stem Vegetables

- Asparagus

- Cardoon

- Celery

- Fennel

- Fiddlehead

- Good King Henry

- Kohlrabi (German turnip)

- Leek

- Nopal

- Prussian Asparagus

- Rhubarb

- Swiss Chard

Fruit Vegetables

Freshly harvested fruits of bell pepper.

Fruit vegetables are collectively one of the many high-value crops that entrepreneurial farmers grow. Consequently, many companies engage in the breeding of superior varieties of vegetable crops such as tomato, eggplant, peppers (Capsicum), bitter gourd, bottle gourd, string beans, and many more.

These vegetables are a common scene in wet market stalls. Likewise, these are commonly grown in residential backyards. Some even utilize the boundary fences as trellises for such climbing plants as string beans or pole sitao, lima beans, and winged bean. Standing trees are likewise used for climbers with relatively large fruits like bottle gourd, luffa and chayote.

Fruit

Fruit is the fleshy or dry ripened ovary of a flowering plant, enclosing the seed or seeds. Thus, apricots, bananas, and grapes, as well as bean pods, corn grains, tomatoes, cucumbers, and (in their shells) acorns and almonds, are all technically fruits. Popularly, however, the term is restricted to the ripened ovaries that are sweet and either succulent or pulpy.

Apricots

Botanically, a fruit is a mature ovary and its associated parts. It usually contains seeds, which have developed from the enclosed ovule after fertilization, although development without fertilization, called parthenocarpy, is known, for example, in bananas. Fertilization induces various changes in a flower: the anthers and stigma wither, the petals drop off, and the sepals may be shed or undergo modifications; the ovary enlarges, and the ovules develop into seeds, each containing an embryo plant. The principal purpose of the fruit is the protection and dissemination of the seed. Fruits are important sources of dietary fibre, vitamins (especially vitamin C), and antioxidants. Although fresh fruits are subject to spoilage, their shelf life can be extended by refrigeration or by the removal of oxygen from their storage or packaging containers. Fruits can be processed into juices, jams, and jellies and preserved by dehydration, canning, fermentation, and pickling. Waxes, such as those from bayberries (wax myrtles), and vegetable ivory from the hard fruits of a South American palm species (Phytelephas macrocarpa) are important fruit-derived products. Various drugs come from fruits, such as morphine from the fruit of the opium poppy.

Types of Fruits

The concept of "fruit" is based on such an odd mixture of practical and theoretical considerations that it accommodates cases in which one flower gives rise to several fruits (larkspur) as well as cases in which several flowers cooperate in producing one fruit (mulberry). Pea and bean plants, exemplifying the simplest situation, show in each flower a single pistil (female structure), traditionally thought of as a megasporophyll or carpel. The carpel is believed to be the evolutionary product of an originally leaflike organ bearing ovules along its margin. This organ was somehow folded along the median line, with a meeting and coalescing of the margins of each half, the result being a miniature closed but hollow pod with one row of ovules along the suture. In many members of the rose and buttercup families, each flower contains a number of similar single-carpelled pistils, separate and distinct, which together represent what is known as an apocarpous gynoecium. In other cases, two to several carpels (still thought of as megasporophylls, although perhaps not always justifiably) are assumed to have fused to produce a single compound gynoecium (pistil), whose basal part, or ovary, may be uniloculate (with one cavity) or pluriloculate (with several compartments), depending on the method of carpel fusion.

Most fruits develop from a single pistil. A fruit resulting from the apocarpous gynoecium (several pistils) of a single flower may be referred to as an aggregate fruit. A multiple fruit represents the gynoecia of several flowers. When additional flower parts, such as the stem axis or floral tube, are retained or participate in fruit formation, as in the apple or strawberry, an accessory fruit results.

Certain plants, mostly cultivated varieties, spontaneously produce fruits in the absence of pollination and fertilization; such natural parthenocarpy leads to seedless fruits such as bananas, oranges, grapes, and cucumbers. Since 1934, seedless fruits of tomato, cucumber, peppers, holly, and others have been obtained for commercial use by administering plant growth substances, such as indoleacetic acid, indolebutyric acid, naphthalene acetic acid, and β-naphthoxyacetic acid, to the ovaries in flowers (induced parthenocarpy).

Seedless watermelon: A seedless watermelon.

Classification systems for mature fruits take into account the number of carpels constituting the original ovary, dehiscence (opening) versus indehiscence, and dryness versus fleshiness. The properties of the ripened ovary wall, or pericarp, which may develop entirely or in part into fleshy, fibrous, or stony tissue, are important. Often three distinct pericarp layers can be identified: the outer (exocarp), the middle (mesocarp), and the inner (endocarp). All purely morphological systems (i.e., classification schemes based on structural features) are artificial. They ignore the fact that fruits can be understood only functionally and dynamically.

Classification of fruits		
	Structure	
Major types	One carpel	Two or more carpels
Dry dehiscent	Follicle—at maturity, the carpel splits down one side, usually the ventral suture; milkweed, columbine, peony, larkspur, marsh marigold.	Capsule—from compound ovary, seeds shed in various ways—e.g., through holes (Papaver—poppies) or longitudinal slits (California poppy) or by means of a lid (pimpernel); flower axis participates in Iris; snapdragons, violets, lilies, and many plant families.
	Legume—dehisces along both dorsal and ventral sutures, forming two valves; most members of the pea family.	Silique—from bicarpellate, compound, superior ovary; pericarp separates as two halves, leaving persistent central septum with seed or seeds attached; dollar plant, mustard, cabbage, rock cress, wall flower.
		Silicle—a short silique; shepherd's purse, pepper grass.
Dry indehiscent	Peanut fruit— (nontypical legume).	Nut—like the achene (see below); derived from 2 or more carpels, pericarp hard or stony; hazelnut, acorn, chestnut, basswood.

	Lomentum—a legume fragmentizing transversely into single-seeded "mericarps"; sensitive plant (Mimosa).	Schizocarp—collectively, the product of a compound ovary fragmentizing at maturity into a number of one-seeded "mericarps"; maple, mallows, members of the mint family (Lamiaceae or Labiatae), geraniums, carrots, dills, fennels.
	Achene—small single-seeded fruit, pericarp relatively thin; seed free in cavity except for its funicular attachment; buttercup, anemones, buckwheat, crowfoot, water plantain.	
	Cypsela—achenelike, but from inferior compound ovary; members of the aster family (Asteraceae or Compositae), sunflowers.	
	Samara—a winged achene; elm, ash, tree-of-heaven, wafer ash.	
	Caryopsis—achenelike; from compound ovary; seed coat fused with pericarp; grass family (Poaceae or Graminae).	
Fleshy (pericarp partly or wholly fleshy or fibrous)	Drupe—mesocarp fleshy, endocarp hard and stony; usually single-seeded; plum, peach, almond, cherry, olive, coconut.	
	Berry—both mesocarp and endocarp fleshy; one-seeded: nutmeg, date; one carpel, several seeds: baneberry, may apple, barberry, Oregon grape; more carpels, several seeds: grape, tomato, potato, asparagus.	
	Pepo—berry with hard rind; squash, cucumber, pumpkin, watermelon.	
	Hesperidium—berry with leathery rind; orange, grapefruit, lemon.	
Structure		
Major types	Two or more carpels of the same flower plus stem axis or floral tube.	Carpels from several flowers plus stem axis or floral tube plus accessory parts.
Fleshy (pericarp partly or wholly fleshy or fibrous)	Pome—accessory fruit from compound inferior ovary; only central part of fruit represents pericarp, with fleshy exocarp and mesocarp and cartilaginous or stony endocarp ("core"); apple, pear, quince, hawthorn, mountain ash.	Multiple fruits—fig (a "syconium"), mulberry, osage orange, pineapple, flowering dogwood.
	Inferior berry—blueberry.	
	Aggregate fleshy fruits—strawberry (achenes borne on fleshy receptacle); blackberry, raspberry (collection of drupelets); magnolia.	

There are two broad categories of fruits: fleshy fruits, in which the pericarp and accessory parts develop into succulent tissues, as in eggplants, oranges, and strawberries; and dry fruits, in which the entire pericarp becomes dry at maturity. Fleshy fruits include (1) the berries, such as tomatoes, blueberries, and cherries, in which the entire pericarp and the accessory parts are succulent tissue, (2) aggregate fruits, such as blackberries and strawberries, which form from a single flower with many pistils, each of which develops into fruitlets, and (3) multiple fruits, such as pineapples and mulberries, which develop from the mature ovaries of an entire inflorescence. Dry fruits include the legumes, cereal grains, capsulate fruits, and nuts.

As strikingly exemplified by the word *nut*, popular terms often do not properly describe the botanical nature of certain fruits. A Brazil nut, for example, is a thick-walled seed enclosed in a likewise thick-walled capsule along with several sister seeds. A coconut is a drupe (a stony-seeded fruit) with a fibrous outer part. A walnut is a drupe in which the pericarp has differentiated into a fleshy outer husk and an inner hard "shell"; the "meat" represents the seed—two large convoluted cotyledons, a minute epicotyl and hypocotyl, and a thin papery seed coat. A peanut is an indehiscent legume fruit. An almond is a drupe "stone"; i.e., the hardened endocarp usually contains a single seed. Botanically speaking, blackberries and raspberries are not true berries but aggregates of tiny drupes. A juniper "berry" is not a fruit at all but the cone of a gymnosperm. A mulberry is a multiple fruit made up of small nutlets surrounded by fleshy sepals. And strawberry represents a much-swollen receptacle (the tip of the flower stalk bearing the flower parts) bearing on its convex surface an aggregation of tiny brown achenes (small single-seeded fruits).

Brazil nut: Hard, indehiscent fruits of the Brazil nut tree (*Bertholletia excelsa*). The fruit on the left has been opened to reveal the large edible seeds in their shells. The tree is found in the Amazonian forests of Brazil, Peru, Colombia, and Ecuador.

Dispersal

Fruits play an important role in the seed dispersal of many plant species. In dehiscent fruits, such as poppy capsules, the seeds are usually dispersed directly from the fruits, which may remain on the plant. In fleshy or indehiscent fruits, the seeds and fruit are commonly moved away from the parent plant together. In many plants, such as grasses and lettuce, the outer integument and ovary wall are completely fused, so seed and fruit form one entity; such seeds and fruits can logically be described together as "dispersal units," or diaspores.

Animal Dispersal

A wide variety of animals' aid in the dispersal of seeds, fruits, and diaspores. Many birds and mammals, ranging in size from mice and kangaroo rats to elephants, act as dispersers when they eat fruits and diaspores. In the tropics, chiropterochory (dispersal by large bats such as flying foxes, *Pteropus*) is particularly important. Fruits adapted to these animals are relatively large and drab in colour with large seeds and a striking (often rank) odour. Such fruits are accessible to bats because of the pagoda-like structure of the tree canopy, fruit placement on the main trunk, or suspension from long stalks that hang free of the foliage. Examples include mangoes, guavas, breadfruit, carob, and several fig species. In South Africa a desert melon (*Cucumis humifructus*) participates in a symbiotic relationship with aardvarks—the animals eat the fruit for its water content and bury their own dung, which contains the seeds, near their burrows.

Additionally, furry terrestrial mammals are the agents most frequently involved in epizoochory, the inadvertent carrying by animals of dispersal units. Burlike fruits, or those diaspores provided with spines, hooks, claws, bristles, barbs, grapples, and prickles, are genuine hitchhikers, clinging tenaciously to their carriers. Their functional shape is achieved in various ways: in cleavers, or goose grass (*Galium aparine*), and in enchanter's nightshade (*Circaea lutetiana*), the hooks are part of the fruit itself; in common agrimony (*Agrimonia eupatoria*), the fruit is covered by a persistent calyx (the sepals, parts of the flower, which remain attached beyond the usual period) equipped with hooks; and in wood avens (*Geum urbanum*), the persistent styles have hooked tips. Other examples are bur marigolds, or beggar's-ticks (*Bidens* species); buffalo bur (*Solanum rostratum*); burdock (*Arctium*); *Acaena*; and many *Medicago* species. The last-named, with dispersal units highly resistant to damage from hot water and certain chemicals (dyes), have achieved wide global distribution through the wool trade. A somewhat different principle is employed by the so-called trample burrs, said to lodge themselves between the hooves of large grazing mammals. Examples are mule grab (*Proboscidea*) and the African grapple plant (*Harpagophytum*). In water burrs, such as those of the water chestnut *Trapa*, the spines should probably be considered as anchoring devices.

Cocklebur: Cocklebur (*Xanthium strumarium*).

Birds, being preening animals, rarely carry burlike diaspores on their bodies. They do, however, transport the very sticky (viscid) fruits of *Pisonia*, a tropical tree of the four-o'clock family, to distant Pacific islands in this way. Small diaspores, such as those of sedges and certain grasses, may also be carried in the mud sticking to waterfowl and terrestrial birds.

Chesnut-mandibled toucan: Chestnut-mandibled, or Swainson's, toucan
(*Ramphastos swainsonii*) consuming a nut.

Synzoochory, deliberate carrying of diaspores by animals, is practiced when birds carry diaspores in their beaks. The European mistle thrush (*Turdus viscivorus*) deposits the viscid seeds of mistletoe (*Viscum album*) on potential host plants when, after a meal of the berries, it whets its bill on branches or simply regurgitates the seeds. The North American (*Phoradendron*) and Australian (*Amyema*) mistletoes are dispersed by various birds, and the comparable tropical species of the plant family Loranthaceae by flower-peckers (of the bird family Dicaeidae), which have a highly specialized gizzard that allows seeds to pass through but retains insects. Plants may also profit from the forgetfulness and sloppy habits of certain nut-eating birds that cache part of their food but neglect to recover everything or that drop units on their way to a hiding place. Best known in this respect are the nutcrackers (*Nucifraga*), which feed largely on the "nuts" of beech, oak, walnut, chestnut, and hazelnut; the jays (*Garrulus*), which hide hazelnuts and acorns; the nuthatches; and the California woodpecker (*Melanerpes formicivorus*), which may embed literally thousands of acorns, almonds, and pecan nuts in bark fissures or holes of trees. Rodents may aid in dispersal by stealing the embedded diaspores and burying them. In Germany, an average jay may transport about 4,600 acorns per season, over distances of up to 4 km (2.5 miles).

Most ornithochores (plants with bird-dispersed seeds) have conspicuous diaspores attractive to such fruit-eating birds as thrushes, pigeons, barbets (members of the bird family Capitonidae), toucans (family Ramphastidae), and hornbills (family Bucerotidae), all of which either excrete or regurgitate the hard part undamaged. Such diaspores have a fleshy, sweet, or oil-containing edible part; a striking colour (often red or orange); no pronounced smell; protection against being eaten prematurely, in the form of acids and tannins that are present only in the green fruit; protection of the seed against digestion, afforded by bitterness, hardness, or the presence of poisonous compounds; permanent attachment; and, finally, absence of a hard outer cover. In contrast to bat-dispersed diaspores, they occupy no special position on the plant. Examples are rose hips, plums, dogwood fruits, barberry, red currant, mulberry, nutmeg fruits, figs, blackberries, and others. The natural and abundant occurrence of *Euonymus*, which is a largely tropical genus, in temperate Europe and Asia, can be understood only in connection with the activities of birds. Birds also contributed substantially to the repopulation with plants of the Krakatoa island group in Indonesia after the catastrophic volcanic eruption there in 1883. Birds have made *Lantana* (originally American) a pest in Indonesia and Australia; the same is true of black cherries (*Prunus serotina*) in parts of Europe, *Rubus* species in Brazil and New Zealand, and olives (*Olea europaea*) in Australia.

Bohemian waxwing (*Bombycilla garrulus*) eating fruit.

Many intact fruits and seeds can serve as fish bait—those of *Sonneratia*, for example, for the catfish *Arius maculatus*. Certain Amazon River fishes react positively to the audible "explosions" of the ripe fruits of *Eperua rubiginosa*. The largest freshwater wetlands in the world, found in Brazil's Pantanal, become inundated with seasonal floods at a time when many plants are releasing their fruits. Pacu fish (*Metynnis*) feed on submerged and floating fruits and disperse the seeds when they defecate. It is thought that at least one plant species (*Bactris glaucescens*) relies exclusively on pacu for seed dispersal.

Fossil evidence indicates that saurochory, dispersal by reptiles, is very ancient. The giant Galapagos tortoise is important for the dispersal of local cacti and tomatoes, and iguanas are known to eat and disperse a number of smaller fruits, including the iguana hackberry (*Celtis iguanaea*). The name alligator apple, for *Annona glabra*, refers to its method of dispersal, an example of saurochory.

Wind Dispersal

Winged fruits are most common in trees and shrubs, such as maple, ash, elm, birch, alder, and dipterocarps (a family of about 600 species of Old World tropical trees). The one-winged propeller type, as found in maple, is called a samara. When fruits have several wings on their sides, rotation may result, as in rhubarb and dock species. Sometimes accessory parts form the wings—for example, the bracts (small green leaflike structures that grow just below flowers) in linden (*Tilia*).

Wind dispersal: Winged fruits of the silver maple (*Acer saccharinum*).

Many fruits form plumes, some derived from persisting and ultimately hairy styles, as in clematis, avens, and anemones; some from the perianth, as in the sedge family (Cyperaceae); and some from the pappus, a calyx structure, as in dandelion and Jack-go-to-bed-at-noon (*Tragopogon*). In woolly fruits and seeds, the pericarp or the seed coat is covered with cottonlike hairs—e.g., willow, poplar or cottonwood, cotton, and balsa. In some cases, the hairs may serve double duty in that they function in water dispersal as well as in wind dispersal.

Salsify: Cluster of plumed fruits on a salsify plant (*Tragopogon porrifolius*).

Poppies have a mechanism in which the wind has to swing the slender fruitstalk back and forth before the seeds are thrown out through pores near the top of the capsule. The inflated indehiscent pods of *Colutea arborea*, a steppe plant, represent balloons capable of limited air travel before they hit the ground and become windblown tumbleweeds.

Other Forms of Dispersal

Geocarpy is defined as either the production of fruits underground, as in the arum lilies (*Stylochiton* and *Biarum*), in which the flowers are already subterranean, or the active burying of fruits by the mother plant, as in the peanut (*Arachis hypogaea*). In the American hog peanut (*Amphicarpa bracteata*), pods of a special type are buried by the plant and are cached by squirrels later on. Kenilworth ivy (*Cymbalaria*), which normally grows on stone or brick walls, stashes its fruits away in crevices after strikingly extending the flower stalks. Not surprisingly, geocarpy is most often encountered in desert plants; however, it also occurs in violet species, in subterranean clover (*Trifolium subterraneum*), and in begonias (*Begonia hypogaea*) of the African rainforest. Barochory, the dispersal of seeds and fruits by gravity alone, is demonstrated by the heavy fruits of horse chestnut.

Vegetable Farming

Vegetable farming is a type of crop production intended primarily for human consumption of the crop's edible parts such as the shoot, leaves, fruits, and roots.

Leaf fruit and root vegetable crops.

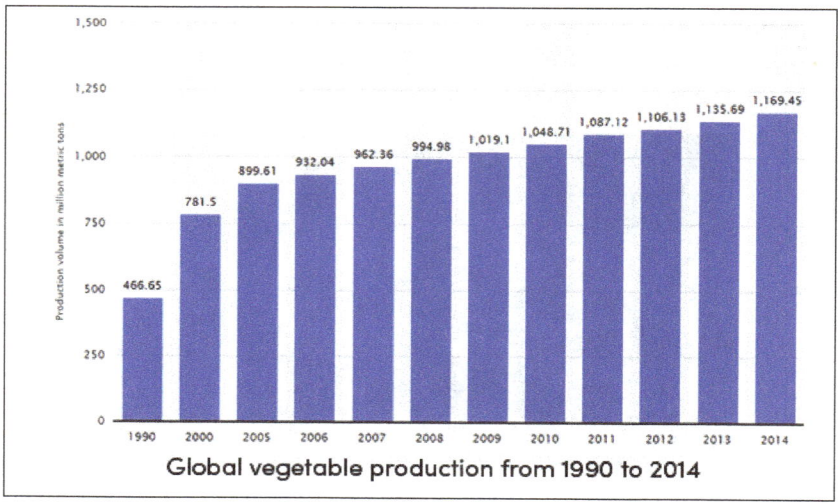

Global vegetable production from 1990 to 2014

Global vegetable crop production has consistently increased in recent years. The growth of vegetable production is shown in the table below.

Despite the fact that vegetable farming is a labor-intensive practice, it's very popular among farmers as a high-income branch of farming. The secret to vegetable farming profitability lies in its high market price of crops, as well as in high demands for vegetables year-round.

Moreover, growing of vegetables is a preferable farm practice in developing and food-insecure countries. Since vegetables are rich in vitamins, minerals, and fibers, they play an important role in diet improvement.

Differences in the Sowing and Planting of Vegetables

Growing of vegetables starts with sowing or planting. Vegetable sowing means putting a seed directly into the prepared soil. On the other hand, vegetable planting includes the practice of putting already grown seedlings into the soil. Seedlings can be grown in the field or in greenhouses. There are two types of seedlings:

- Bare root seedlings are grown in the field from the seed. As the name implies, their roots are separated from the soil when they are moved to the planting site.

- Root ball seedlings are grown in pots or blocks and moved to the planting site with the soil attached to the roots.

The difference between sowing and planting of vegetables.

Sowing is a recommended practice for vegetables with delicate roots and taproots, such as carrots, turnips, and radishes. Growing of vegetables from seedlings is more appropriate in cases such as slow-growing perennials, crops with fine and expensive seed, and warm-season crops. Planting is recommended for annual vegetable crops when the soil is too cold or moist for direct sowing. For example, onion and asparagus are vegetables that are usually planted.

Types of Production

Vegetable production operations range from small patches of crops, producing a few vegetables for family use or marketing, to the great, highly organized and mechanized farms common in the most technologically advanced countries.

In technologically developed countries the three main types of vegetable farming are based on production of vegetables for the fresh market, for canning, freezing, dehydration, and pickling, and to obtain seeds for planting.

Production for the Fresh Market

This type of vegetable farming is normally divided into home gardening, market gardening, truck farming, and vegetable forcing.

Home gardening provides vegetables exclusively for family use. About one-fourth of an acre (one-tenth of a hectare) of land is required to supply a family of six. The most suitable vegetables are those producing a large yield per unit of area. Bean, cabbage, carrot, leek, lettuce, onion, parsley, pea, pepper, radish, spinach, and tomato are desirable home garden crops.

Market gardening produces assorted vegetables for a local market. The development of good roads and of motor trucks has rapidly extended available markets; the market gardener, no longer forced to confine his operations to his local market, often is able to specialize in the production of a few, rather than an assortment, of vegetables; a transformation that provides the basis for a distinction between market and truck gardening in the mid-20th century. Truck gardens produce specific vegetables in relatively large quantities for distant markets.

In the method known as forcing, vegetables are produced out of their normal season of outdoor production under forcing structures that admit light and induce favourable environmental conditions for plant growth. Greenhouses, cold frames, and hotbeds are common structures used. Hydroponics, sometimes called soilless culture, allows the grower to practice automatic watering and fertilizing, thus reducing the cost of labour. To successfully compete with other fresh market producers, greenhouse vegetable growers must either produce crops when the outdoor supply is limited or produce quality products commanding premium prices.

Production for Processing

Processed vegetables include canned, frozen, dehydrated, and pickled products. The cost of production per unit area of land and per ton is usually less for processing crops than for the same crops grown for market because raw material appearance is not a major quality factor in processing. This difference allows lower land value, less hand labour, and lower handling cost. Although many kinds of vegetables can be processed, there are marked varietal differences within each species in adaptability to a given method.

Specifications for vegetables for canning and freezing usually include small size, high quality, and uniformity. For many kinds of vegetables, a series of varieties having different dates of maturity is required to ensure a constant supply of raw material, thus enabling the factory to operate with an even flow of input over a long period. Acceptable processed vegetables should have a taste, odour, and appearance comparable with the fresh product, retain nutritive values, and have good storage stability.

Vegetables raised for Seed Production

This type of vegetable farming requires special skills and techniques. The crop is not ready for harvest when the edible portion of the plant reaches the stage of maturity; it must be carried through further stages of growth. Production under isolated conditions ensures the purity of seed yield. Special techniques are applied during the stage of flowering and seed development and also in harvesting and threshing the seeds.

Production Factors and Techniques

Profitable vegetable farming requires attention to all production operations, including insect, disease, and weed control and efficient marketing. The kind of vegetable grown is mainly determined by consumer demands, which can be defined in terms of variety, size, tenderness, flavour, freshness, and type of pack. Effective management involves the adoption of techniques resulting in a steady flow of the desired amount of produce over the whole of the natural growing season of the crop. Many vegetables can be grown throughout the year in some climates, although yield per acre for a given kind of vegetable varies according to the growing season and region where the crop is produced.

Climate

Climate involves the temperature, moisture, daylight, and wind conditions of a specific region. Climatic factors strongly affect all stages and processes of plant growth.

Temperature

Temperature requirements are based on the minimum, optimum, and maximum temperatures during both day and night throughout the period of plant growth. Requirements vary according to the type and variety of the specific crop. Based on their optimum temperature ranges, vegetables may be classed as cool-season or warm-season types. Cool-season vegetables thrive in areas where the mean daily temperature does not rise above 70° F (21° C). This group includes the artichoke, beet, broccoli, brussels sprouts, cabbage, carrot, cauliflower, celery, garlic, leek, lettuce, onion, parsley, pea, potato, radish, spinach, and turnip. Warm-season vegetables, requiring mean daily temperature of 70° F or above, are intolerant of frost. These include the bean, cucumber, eggplant, lima bean, okra, muskmelon, pepper, squash, sweet corn (maize), sweet potato, tomato, and watermelon.

Premature seeding, or bolting, is an undesirable condition that is sometimes seen in fields of cabbage, celery, lettuce, onion, and spinach. The condition occurs when the plant goes into the seeding stage before the edible portion reaches a marketable size. Bolting is attributed to either extremely low or high temperature conditions in combination with inherited traits. Specific vegetable strains or varieties may exhibit significant differences in their tendency to bolt.

Young cabbage or onion plants of relatively large size may bolt upon exposure to low temperatures near 50° to 55° F (10° to 13° C). At high temperatures of 70° to 80° F (21° to 27° C) lettuce plants do not form heads and will show premature seeding. The fruit sets of tomatoes are adversely affected by relatively low and relatively high temperatures. Tomato breeders, however, have developed several new varieties, some setting fruits at a temperature as low as 40° F (4° C) and others at a temperature as high as 90° F (32° C).

Moisture

The amount and annual distribution of rainfall in a region, especially during certain periods of development, affects local crops. Irrigation may be required to compensate for insufficient rainfall.

For optimum growth and development, plants require soil that supplies water as well as nutrients dissolved in water. Root growth determines the extent of a plant's ability to absorb water and nutrients, and in dry soil root growth is greatly retarded. Extremely wet soil also retards root growth by restricting aeration. Atmospheric humidity, the moisture content of the air, also contributes moisture. Certain seacoast areas characterized by high humidity are considered especially adapted to the production of such crops as the artichoke and lima bean. High humidity, however, also creates conditions favourable for the development of certain plant diseases.

Daylight

Light is the source of energy for plants. The response of plants to light is dependent upon light intensity, quality, and daily duration, or photoperiod. The seasonal variation in day length affects the growth and flowering of certain vegetable crops. Continuation of vegetative growth, rather than early flower formation, is desirable in such crops as spinach and lettuce. When planted very late in the spring, these crops tend to produce flowers and seeds during the long days of summer before they attain sufficient vegetative growth to produce maximum yields. The minimum photoperiod required for formation of bulbs in garlic and onion plants differs among varieties, and local day length is a determining factor in the selection of varieties.

Each of the climatic factors affects plant growth, and can be a limiting factor in plant development. Unless each factor is of optimum quantity or quality, plants do not achieve maximum growth. In addition to the importance of individual climatic factors, the interrelationship of all environmental factors affects growth.

Certain combinations may exert specific effects. Lettuce usually forms a seedstalk during the long days of summer, but the appearance of flowers may be delayed, or even prevented, by relatively low temperature. An unfavourable temperature combined with unfavourable moisture conditions may cause the dropping of the buds, flowers, and small fruits of the pepper, reducing the crop yield. Desirable areas for muskmelon production are characterized by low humidity combined with high temperature. In the production of seeds of many kinds of vegetables, absence of rain, or relatively light rainfall, and low humidity during ripening, harvesting, and curing of the seeds are very important.

Site

The choice of a site involves such factors as soil and climatic region. In addition, with the continued trend toward specialization and mechanization, relatively large areas are required for commercial production, and adequate water supply and transportation facilities are essential. Topography—that is, the surface of the soil and its relation to other areas—influences efficiency of operation. In modern mechanized farming, large, relatively level fields allow for lower operating costs. Power equipment may be used to modify topography, but the cost of such land renovation may be prohibitive. The amount of slope influences the type of culture possible. Fields with a moderate slope should be contoured; a process that may involve added expense for the building of terraces and diversion ditches. The direction of a slope may influence the maturation time of a crop or may result in drought, winter injury, or wind damage. A level site is generally most desirable, although a slight slope may assist drainage. Exposed sites are not suitable for vegetable farming because of the risk of damage to plants by strong winds.

The soil stores mineral nutrients and water used by plants, as well as housing their roots. There are two general kinds of soils—mineral and the organic type called muck or peat. Mineral soils include sandy, loamy, and clayey types. Sandy and loamy soils are usually preferred for vegetable production. Soil reaction and degree of fertility can be determined by chemical analysis. The reaction of the soil determines to a great extent the availability of most plant nutrients. The degree of acid, alkaline, or neutral reaction of a soil is expressed as the pH, with a pH of 7 being neutral, points below 7 being acid, and those above 7 being alkaline. The optimum pH range for plant growth varies from one crop to another. A soil can be made more acid, or less alkaline, by applying an acid-producing chemical fertilizer such as ammonium sulfate.

The inherent fertility of soils affects production quantity, and a sound fertility program is required to maintain productivity. The ability of a soil to support plant life and produce abundant harvests is dependent on the immediately available nutrients in the soil and on the rate of release of additional nutrients that are present but not available to plants. The rate of release of these additional nutrients is affected by such factors as microbial action, soil temperature, soil moisture, and aeration. Depletion of soil fertility may occur as a result of crop removal, erosion, leaching, and volatilization, or evaporation, of nutrients.

Soil Preparation and Management

Soil preparation for vegetable growing involves many of the usual operations required for other crops. Good drainage is especially important for early vegetables because wet soil retards development. Sands are valuable in growing early vegetables because they are more readily drained than the heavier soils. Soil drainage accomplished by means of ditches or tiles is more desirable than the drainage obtained by planting crops on ridges because the former not only removes the excess water but also allows air to enter the soil. Air is essential to the growth of crop plants and to certain beneficial soil organisms making nutrients available to the plants.

When crops are grown in succession, soil rarely needs to be plowed more than once each year. Plowing incorporates sod, green-manure crops, and crop residues in the soil; destroys weeds and insects; and improves soil texture and aeration. Soils for vegetables should be fairly deep. A depth of six to eight inches (15 to 20 centimetres) is sufficient in most soils.

Soil management involves the exercise of human judgment in the application of available knowledge of crop production, soil conservation, and economics. Management should be directed toward producing the desired crops with a minimum of labour. Control of soil erosion, maintenance of soil organic matter, the adoption of crop rotation, and clean culture are considered important soil-management practices.

Soil erosion, caused by water and wind, is a problem in many vegetable-growing regions because the topsoil is usually the richest in fertility and organic matter. Soil erosion by water can be controlled by various methods. Terracing divides the land into separate drainage areas, with each area having its own waterway above the terrace. The terrace holds the water on the land, allowing it to soak into the soil and reducing or preventing gullying. In the contouring system, crops are planted in rows at the same level across the field. Cultivation proceeds along the rows rather than up and down the hill. Strip cropping consists of growing crops in narrow strips across a slope, usually on the contour. Soil erosion by wind can be controlled by the use of windbreaks of various kinds, by keeping the soil well supplied with humus, and by growing cover crops to hold the soil when the land is not occupied by other crops.

Maintenance of the organic-matter content of the soil is essential. Organic matter is a source of plant nutrients and is valuable for its effect on certain properties of the soil. Loss of organic matter is the result of the action of micro-organisms that gradually decompose it to carbon dioxide. The addition of manures and the growing of soil-improving crops are efficient means of supplying soil organic matter. Soil-improving crops are grown solely for the purpose of preparing the soil for the growth of succeeding crops. Green-manure crops, grown especially for soil improvement, are turned under while still green and usually are grown during the same season of the year as the vegetable crops. Cover crops, raised for both soil protection and improvement, are only grown during seasons when vegetable crops do not occupy the land. When a soil-improving crop is turned under, the various nutrients that have contributed to the growth of the crop are returned to the soil, adding a quantity of organic matter. Both legumes, those plants such as peas and beans having fruits and seeds formed in pods, and nonlegumes are effective soil-improving crops. The legumes, however, are more valuable, because they contribute nitrogen as well as humus. The rate of decomposition of plant material depends on the kind of crop, its stage of growth, and soil temperature and moisture. The more succulent the material is at the time it is turned under, the more quickly it decomposes. Because dry material decomposes more slowly than green material, it is desirable to turn under soil-improving crops before they are mature, unless considerable time is to elapse between the plowing and the planting of the succeeding crop. Plant material decomposes most rapidly when the soil is warm and well supplied with moisture. If soil is dry when a soil-improving crop is turned under, little or no decomposition will occur until rain or irrigation supplies the necessary moisture.

The chief benefits derived from crop rotation are the control of disease and insects and the better use of the resources of the soil. Rotation is a systematic arrangement for the growing of different crops in a more or less regular sequence on the same land. It differs from succession cropping in that rotation cropping covers a period of two, three, or more years, while in succession cropping two or more crops are grown on the same land in one year. In many regions vegetable crops are grown in rotation with other farm crops. Most vegetables grown as annual crops fit into a four-or five-year rotation plan. The system of intercropping, or companion cropping, involves the growing of two or more kinds of vegetables on the same land in the same growing season. One of the vegetables must be a small-growing and quick-maturing crop; the other must be larger and late maturing.

In the practice of clean culture, commonly followed in vegetable growing, the soil is kept free of all competing plants through frequent cultivation and the use of protective coverings, or mulches, and weed killers. In a clean vegetable field the possibility of attack by insects and disease-incitant organisms, for which plant weeds serve as hosts, is reduced.

Propagation

Propagation of crop plants, involving the formation and development of new individuals in the establishment of new plantings, is usually accomplished by the use of either seeds or the vegetative parts of plants. The first type, known as sexual propagation, is used for asparagus, bean, broccoli, cabbage, carrot, cauliflower, celery, cucumber, eggplant, leek, lettuce, lima bean, okra, onion, muskmelon, parsley, pea, pepper, pumpkin, radish, spinach, sweet corn (maize), squash, tomato, turnip, and watermelon. The second type, asexual propagation, is used for the artichoke, garlic, girasole, potato, rhubarb, and sweet potato.

Although seed cost is a small portion of the total cost of crop production, seed quality strongly affects crop success or failure. Good seed should be accurately labelled, clean, graded to size, viable, and free of diseases and insects. The reliability of the seed house is an important factor in obtaining good-quality seed. Viability, or ability to grow, and longevity, the period of viability, are characteristics of seeds of any vegetable kind. In cool, dry storage conditions, those vegetable seeds having comparatively short longevity of one to two years are okra, onion, parsley, and sweet corn. Seeds having three-year longevity are those of the asparagus, bean, carrot, leek, and pea; four-year longevity is characteristic of the beet, chard, pepper, pumpkin, and tomato seeds; longevity of five years characterizes the seeds of broccoli, cabbage, cauliflower, celery, cucumber, eggplant, lettuce, muskmelon, radish, spinach, squash, turnip, and watermelon. The dry seeds of all vegetables, when packed under vacuum in hermetically sealed cans, should remain viable for a longer period than seeds stored under less protective conditions.

Crops grown from hybrid seeds (the offspring of two or more selected parental varieties and known as F_1) yield vegetables of high quantity and quality. The hybrid-seed industry is based on the production of new seed each year from the controlled pollination of selected parents found to produce the desired combination of characters in the progeny. In the early 1980s the number of F_1 hybrids was increasing in Japan, the United States, and other technically advanced countries. The number of F_1 hybrids varied with the kind of vegetable, but none had yet been introduced for the bean, celery, lettuce, okra, parsley, or pea.

Planting

Most vegetable crops are planted in the field where they are to grow to maturity. A few kinds are commonly started in a seedbed, established in the greenhouse or in the open, and transplanted as seedlings. Asparagus seeds are planted in a seedbed to produce crowns used for field setting. Some vegetables can be either directly seeded in the field or grown from transplants. These include broccoli, cabbage, cauliflower, celery, eggplant, leek, lettuce, onion, pepper, and tomato. The time and method of planting seeds and plants of a particular vegetable influence the success or failure of the crop. Important factors include the depth of planting, the rate of planting, and the spacing both between rows and between plants within a row.

Factors to be considered in determining the time of planting include soil and weather conditions, kind of crop, and desired harvest time. When more than one planting of a crop is made, the second and later plantings should be timed to provide a continuous harvest for the period desired. The soil temperature required for germination of the planted seed varies markedly with the various kinds of vegetables. Vegetables that will not germinate at a temperature below 60°F (16 °C) include the bean, cucumber, eggplant, lima bean, muskmelon, okra, pepper, pumpkin, squash, and watermelon. Temperatures higher than 90°F (32 °C) are not favourable for the germination of seeds of celery, lettuce, lima bean, parsley, pea, and spinach.

The quantity of seeds planted, or rate of planting, is mainly determined by the characteristics of the vegetable plant. The size of seeds affects the number of plants raised in a given area. Watermelon varieties, for example, differ in seed size expressed as weight. The Sugar Baby variety has an average weight of 1.4 ounces (41 grams) for 1,000 seeds; those of Blackstone variety average 4.4 ounces (125 grams). If the two are grown on two separate plots of the same area and 4.4 ounces of seeds of each cultivar are planted, the result would be three times as many of the Sugar Baby plants as the Blackstone type. Seed size and plant-growth pattern of a vegetable are major factors that govern the number of plants raised in a given area. The trend in the early 1980s was to increase plant population for many crops to achieve the greatest yield possible without impairing quality. As plant population increases per unit area, a point is reached at which each plant begins to compete for certain essential growth factors—e.g., nutrients, moisture, and light. When the population is below the level in which competition between plants occurs, increased population will have no effect on individual plant performance, and the yield per unit area will increase in direct proportion to the increment of population. When competition for essential growth factors occurs, however, yield per plant decreases.

Early harvest and economical use of space are the principal objectives of growing vegetable crops from transplants produced in a greenhouse or outdoor seedbed. It is easier to care for young plants of the cabbage, cauliflower, celery, onion, and tomato in small seedbeds than to sow the seeds in the place where the crop is to grow and mature. Land is free longer for another crop, and weeds, insects, diseases, and irrigation are more readily and economically controlled. The production of transplants is often a specialty of growers who sell their produce to other vegetable growers. The seeds may be planted at a rate three to six times that commonly used for a direct-seeded field. The young plants are removed for use as transplants when they reach the desired size and age, approximately 40 to 60 days after seeding.

Care of Crops during Growth

Practices required for a vegetable crop growing in the field include cultivation; irrigation; application of fertilizers; control of weeds, diseases, and insects; protection against frost; and the application of growth regulators if necessary.

Cultivation

Cultivation refers to stirring the soil between rows of vegetable plants. Because weed control is the most important function of cultivation, this work should be performed at the most favourable time for weed killing, when the weeds are breaking through the soil surface. When the plants are grown on ridges, it is necessary to cover the basal plant portion with soil in the case of such vegetables as asparagus, carrot, garlic, leek, onion, potato, sweet corn, and sweet potato.

Irrigation

Vegetable production requires irrigation in arid and semi-arid regions, and irrigation is frequently used as insurance against drought in more humid regions. In areas having intermittent rain for five or six months, with little or none during the remainder of the year, irrigation is essential throughout the dry season and may also be needed between rainfalls in the rainy season. The two types of land irrigation generally suited to vegetables are surface irrigation and sprinkler irrigation. A level site is required for surface irrigation, in which the water is conveyed directly over the field in open ditches at a slow, nonerosive velocity. Where water is scarce, pipelines may be used, eliminating losses caused by seepage and evaporation. The distribution of water is accomplished by various control structures, and the furrow method of surface irrigation is frequently employed because most vegetable crops are grown in rows. Sprinkler irrigation conveys water through pipes for distribution under pressure as simulated rain.

Irrigation requirements are determined by both soil and plant factors. Soil factors include texture, structure, water-holding capacity, fertility, salinity, aeration, drainage, and temperature. Plant factors include type of vegetable, density and depth of the root system, stage of growth, drought tolerance, and plant population.

Fertilizer Application

Soil fertility is the capacity of the soil to supply the nutrients necessary for good crop production, and fertilizing is the addition of nutrients to the soil. Chemical fertilizers may be used to supply the needed nitrogen, phosphorus, and potassium. Chemical tests of soil, plant, or both are used to determine fertilizer needs, and the rate of application is usually based on the fertility of the soil, the cropping system employed, the kind of vegetable to be grown, and the financial return that might be expected from the crop. Methods of fertilizer application include scattering and mixing with the soil before planting; application with a drill below the surface of the soil at the time of planting; row application before or at planting time; and row application during plant growth, also called side-dressing. Plowed down broadcast fertilizers have recently been used in combination with high analysis liquid fertilizers applied at planting or as a side-dressed band. Mechanical planting devices may employ fertilizer attachments to plant the fertilizer in the form of bands near the seed. For most vegetables, the bands are placed from two to three inches (five to 7.5 centimetres) from the seed, either at the same depth or slightly below the seed.

Weed Control

Weeds (plants growing where they are not wanted) reduce crop yield, increase production cost, and may harbour insects and diseases that attack crop plants. Methods employed to control weeds include hand weeding, mechanical cultivation, application of chemicals acting as herbicides, and a combination of mechanical and chemical means. Herbicides, selective chemical weed killers, are absorbed by the plant and induce a toxic reaction. The amount and type of herbicide that can be safely used to protect vegetable crops depends on the tolerance of the specific crops to the chemical. Most herbicides are applied as a spray, and the appropriate time for application is determined by the composition of the herbicide and the kind of vegetable crop to be treated. Preplanting treatments are applied before the crop is planted; preemergence treatments are applied after the crop is planted but before its seedlings emerge from the soil; and postemergence treatments are applied to the growing crop at a definite stage of growth.

Disease and Insect Control

The production of satisfactory crops requires rigorous disease- and insect-control measures. Crop yield may be lowered by disease or insect attack, and when plants are attacked at an early stage of growth the entire crop may be lost. Reduction in the quality of vegetable crops may also be caused by diseases and insects. Grades and standards for market vegetables usually specify strict limits on the amount of disease and insect injury that may be present on vegetables in a designated grade.

Vegetables remain vulnerable to insect and disease damage after harvesting, during the marketing and handling processes. When a particular plant pest is identified, the grower can select and apply appropriate control measures. Application of insect control at the times specific insects usually appear or when the first insects are noticed is usually most effective. Effective disease control usually requires preventive procedures.

Diseases are incited by such living organisms as bacteria, fungi, and viruses. Harmful material enters the plant, develops during an incubation period, and finally causes infection, the reaction of the plant to the pathogen, or disease-producing organism. Control is possible during the inoculation and incubation phases, but when the plant reaches the infection stage it is already damaged. Typical plant diseases include mildew, leaf spots, rust, and wilt. Chemical fungicides may be used to control disease, but the use of disease-resistant plant varieties is the most effective means of control.

Vegetable breeders have developed plant varieties resistant to one or more diseases; such varieties are available for the bean, cabbage, cucumber, lettuce, muskmelon, onion, pea, pepper, potato, spinach, tomato, and watermelon.

Insects are usually controlled by the use of chemical insecticides that kill through toxic action. Many insecticides are toxic to harmful insects but do not affect bees, which are valuable for their role in pollination.

Frost Protection

Frost protection may be accomplished by increasing the amount of heat radiated from the soil when frost is likely to occur. Irrigation on the day before a predicted frost provides additional moisture in the soil to increase the amount of heat given off as infrared rays. This extra heat protects the plants from frost injury. A continuous supply of water provided by sprinkler irrigation may also protect plants from frost. As the water freezes on the plant leaves, it loses heat that is absorbed by the plant leaves, maintaining leaf temperature at 32° F (0° C). Because of the sugars and other substances in plant cells, the freezing point of cell sap is somewhat lower than 32° F.

Growth Regulators

It is sometimes desirable to retard or accelerate maturity in vegetable crops. A chemical compound may be applied to prevent sprouting in onion crops. It is applied in the field sufficiently early for absorption by the still-green foliage but late enough to avoid suppressing the bulb yield. Another substance may be used to end the dormancy, or rest period, of newly harvested potato tubers intended for planting. The treated seed potatoes have uniform sprout emergence. The same substance is applied to celery from two to three weeks before harvest to elongate the stalks and increase the yield and is also used to accelerate maturity in artichokes. A chemical compound, applied when adverse weather conditions prevail during the period of fruit setting, has been used to encourage fruit set.

Harvesting

The stage of development of vegetables when harvested affects the quality of the product reaching the consumer. In some vegetables, such as the bean and pea, optimum quality is reached well in advance of full maturity and then deteriorates, although yield continues to increase. Factors determining the harvest date include the genetic constitution of the vegetable variety, the planting date, and environmental conditions during the growing season. Successive harvest dates may be obtained either by planting varieties having different maturity dates or by changing the sequence of planting dates of one particular variety. The successive method is applicable to such crops as broccoli, cabbage, cauliflower, muskmelon, onion, pea, sweet corn (maize), tomato, and watermelon. Certain varieties of the carrot, celery, cucumber, lettuce, parsley, radish, spinach, or summer squash can be sown in succession throughout most of the year in some climates, thus prolonging the harvest period.

Hand harvesting is employed along with various mechanical aids for broccoli, cabbage, cauliflower, muskmelon, and pepper crops. Many vegetables grown for processing and some vegetables destined for the fresh market are mechanically harvested. Harvesting operations may be performed by a single machine in a single step for such vegetable crops as the bean, beet, carrot, lima bean, onion, pea, potato, radish, spinach, sweet corn, sweet potato, and tomato. Designers of harvesting machinery have been working to develop a multiple-picking harvester capable of adjustment for use with more than one crop. Vegetable breeders have been able to produce vegetables with characteristics suitable for machine harvesting, including compact plant growth, uniform development, and concentrated maturity.

Storage

Fresh vegetables are living organisms, and there is a continuation of life processes in the vegetable after harvest. Changes that occur in the harvested, nonprocessed vegetable include water loss, conversion of starches to sugars, conversion of sugars to starches, flavour changes, colour changes, toughening, vitamin gain or loss, sprouting, rooting, softening, and decay.

Some changes result in quality deterioration; others improve quality in those vegetables that complete ripening after harvest. Postharvest changes are influenced by such factors as kind of crop, air temperature and circulation, oxygen and carbon dioxide contents and relative humidity of the atmosphere, and disease-incitant organisms. To maintain the fresh vegetable in the living state, it is usually necessary to slow the life processes, though avoiding death of the tissues, which produces gross deterioration and drastic differences in flavour, texture, and appearance.

Storage of vegetables contributes to price stabilization by carrying over produce from periods of high production to periods of low production. It also extends the period of consumption of many kinds of vegetables. Storage conditions can contribute to the preservation of the natural living state of the edible portion and to the prevention of deterioration through control of temperature, relative humidity, and the quality of the produce to be stored. Vegetables for storage must be free from mechanical, insect, and disease injury and should be at the proper stage of maturity.

Common (unrefrigerated) storage and cold (refrigerated) storage are the methods generally employed for vegetables. Common storage, lacking precise control of temperature and humidity, includes the use of insulated storage houses, outdoor cellars, or mounds. Cold storage allows precise regulation of temperature and humidity and maintenance of constant conditions by use of a refrigeration and ventilation system. Temporary storage, suitable only for very brief storage periods, is frequently practiced in the shipping season when large lots are accumulated for carload or truck quantities. The refrigerator car or truck is a means of temporary storage used to protect produce while it is in transit. Short-term storage may last for four or six weeks. Economic factors, such as the probability that prices will increase later in the season, encourage long-term storage of such perishable vegetables as the onion, potato, and sweet potato.

Premarketing Operations and Selling

Premarketing operations include washing, trimming, waxing, precooling, grading, prepackaging, and packaging. Vegetables often require washing after harvest to remove any adhering soil particles. Such vegetables as the beet, carrot, celery, lettuce, radish, spinach, and turnip are trimmed before washing to remove discoloured leaves or to cut back the green tops. Waxing of the cucumber, muskmelon, pepper, potato, sweet potato, and tomato gives the product a bright appearance and controls shrivelling through reduction of moisture loss.

Precooling

Precooling, the rapid removal of heat from freshly harvested vegetables, allows the grower to harvest produce at optimum maturity with greater assurance that it will reach the consumer at maximum quality. Precooling benefits the vegetable by slowing the natural deterioration that starts shortly after harvest, slowing the growth of decay organisms and reducing wilt by retarding water loss. The major precooling methods include hydrocooling, contact icing, vacuum cooling, and air cooling. In hydrocooling the vegetable is cooled by direct contact with cold water flowing through the packed containers and absorbing heat directly from the produce. In contact icing crushed ice is placed in the package or spread over a stack of packages to precool the contents. The vacuum cooling process produces rapid evaporation of a small quantity of water, lowering the temperature of the crop to the desired level. Air cooling involves the exposure of vegetables to cold air; the air must be as cold as possible for rapid cooling but not low enough to freeze the produce exposed to the direct air blast.

The preferred method of precooling varies according to the physical characteristics of the vegetable. Hydrocooling is recommended for the asparagus, beet, broccoli, carrot, cauliflower, celery, muskmelon, pea, radish, summer squash, and sweet corn (maize); cabbage, lettuce, and spinach are suited to vacuum cooling; air cooling is preferred for bean, cucumber, eggplant, pepper, and tomato. After the produce is precooled, it is desirable to maintain low temperature by shipping in refrigerator cars or trucks, by storing in cold-storage rooms, and by refrigeration in retail stores.

Grading

Uniformity in size, shape, colour, and ripeness is of great importance in marketing any vegetable product, and can be secured through grading. The establishment of standard grades furnishes a basis of trade. Grade standards are based mainly on general appearance, size, trueness to type, and freedom from blemishes and defects.

Packaging

Prepackaging, or consumer packaging, has become a highly organized practice, often employing elaborate equipment. The product is placed in bags made of transparent film, trays or cartons overwrapped with transparent film, or mesh or paper bags. The packaging of produce in consumer packages lends itself to self-service in retail stores. The production region is often the most satisfactory location for prepackaging, especially when a packaging centre serves a large vegetable-growing area.

Master containers for consumer packages are commonly made of paperboard. Cartons, bags, baskets, boxes, crates, and hampers of various kinds and sizes are all used in packaging vegetables for marketing. The type of container is selected to fit the kind of vegetable; it furnishes a convenient means for transport, loading, and stacking, with security and economy of space. Uniform product throughout the package is an important consideration in packing vegetables.

Selling

Producers sell vegetables through various retail and wholesale practices. Retail sales are made directly to the consumer, often through roadside stands. Many growers sell most of their produce at wholesale to retail stores, to various types of buyers on local markets in nearby cities, or in regional markets. Growers located long distances from markets sell largely to wholesale dealers or jobbers.

Some growers have contracts with processors. Wholesale marketing arrangements are also made through auction markets in the producing regions and through cooperative organizations of producers.

Fruit Farming

Tropical and sub-tropical fruit trees are generally evergreen and perennial and are frost sensitive with little growth below 10 °C. Tropical species are distinct from subtropical species in that they need humid conditions and are sensitive to temperatures below 20 °C. They thrive in climates where average mean temperature ranges are higher than 10 °C for the coldest month. The widely cultivated and globally traded tropical and subtropical fruit trees are mango, various species of citrus, avocado, papaya, and kiwi.

Scope and Importance of Fruit Crops

Fruit farming is one of the important and age-old practices; Cultivation of fruits plays an important role in the overall status of mankind and the nation. The standard of living of the people of a country is depending upon the production and per capita consumption of several fruits. Fruit growing also has more economic advantages.

Profitable Fruit Farming

- High yield: First of all, you can earn a better return than many of the field crops from a well maintained and established fruit garden. From a unit area, you can realize more agronomic fruit crops.

- High net profits: Though the initial cost of establishment of an orchard is height and it is compensated by higher productivity or due to the high value of produce.

- Efficient utilization of resources: Agronomic fruit crops are seasonal and hence, farmers have to engage themselves in other occupations during slack seasons. However, fruit cultivation is being perennial. Then, it enables the grower to remain engaged throughout the year in farm operations and utilize full resources and assets like machinery, farm, land, water, etc.

- Foreign trade potential: Many fresh fruits and processed products and spices have excellent export potential. Farmers can earn an excellent amount of foreign money from international trade.

Continuous cash flow:

- The harvesting process of most of the fruit is continuous. Due to the perishable nature, it wants to be marketed immediately after harvest. This gives a source of a continuous flow of inputs and other expenses of immediate nature as agronomic crops that are harvested one time.

Economic Importance of Fruit Farming

- High productivity: From a unit area of land more crop yield is realized from fruit crops than any of the agronomic crops. The average yields of Papaya, Banana and also Grapes are 10 to 15 times than that of agronomic crops.

- High net profit: Through, the initial cost of establishment of an orchard is high, and it is compensated by higher net profit due to higher productivity or high value of produce.

- Efficient utilization of resources: Growing of fruits being perennial, and enables the grower to remain engaged throughout the year in farm operations. And to utilize fully the resources and assets such as machinery, labor, and land water for production purposes throughout the year compared to agronomic crops.

Cultivation

Site Selection

The site of a fruit-growing enterprise is as significant in determining its success as the varieties grown. In fact, variety and site together set a ceiling on the productivity and profit that can be realized under the best management. In most developed fruit regions microclimatic conditions (climate at plant height, as influenced by slight differences in soil, soil covering, and elevation) and soil conditions are the two components of a site that determine its desirability for a fruit-growing enterprise. Sometimes (particularly with highly perishable fruits) transportation to market must also be considered.

Local conditions at a site that expose it to unusual frost hazard are as detrimental to citrus in Florida as they are to peach trees in New Zealand and apple trees in the south of England. In regions and sites where temperatures during the season may drop no more than a few degrees below freezing, artificial frost protection is sometimes used. This is accomplished by open-flame burning (petroleum bricks, logs, etc.) or heating of metal objects with oil, gas, propane, electricity, etc. (stones or stacks that radiate heat). Another technique is the spraying of water on plants (e.g., strawberries) as long as the temperature is below freezing.

For highest productivity, most fruit trees must root extensively to a depth of three feet (one metre) or more. Heavy subsoil or other conditions causing imperfect internal drainage may result in shallow, weak root systems that do not take water and nutrients efficiently from the soil. In semi-arid and arid regions, accumulation of saline soils in a subsurface layer sometimes limits rooting of fruit trees, causes abnormal foliar symptoms, and reduces yields. Tiling and surface ditching help decrease water accumulation in poorly drained subsoils and reduce wet spots in otherwise satisfactory sites. Special control of irrigation procedures and periodic leaching may alleviate the worst salt effects in saline soils. Choice of tolerant species, varieties, and rootstocks may make fruit growing economical on imperfectly drained or mildly saline sites, though plants rarely perform as well as they do on sites free from these difficulties. Coconuts, however, tolerate saline soil conditions near tropical saltwater coasts.

Once selected, a site is cleared, levelled (if needed), and cultivated. Then drainage, irrigation, and road systems are installed as required. In rolling or sloping terrain, where contour planting is needed to control erosion and conserve moisture, the locations of the plant or row positions are determined by the contour terraces and waterways established. In old lands, nematode or other pest populations make fumigation necessary before planting. In some problem California soils, giant plows and treaded tractors turn the soil to depths of three to six feet (one to two metres). In very infertile sites, or sites where the physical condition of the surface soil is poor, it may be helpful to grow a succession of leguminous cover crops for a year or more before planting and/or apply a fertilizer containing major fertilizer elements (nitrogen, potassium, phosphorus, calcium, sulfur) and all or certain trace elements (iron, manganese, boron, zinc, copper, molybdenum) and lime, based on a soil test.

Planting and Spacing Systems

Growth, flowering habits, and light requirements on the one hand, and management problems on the other, determine the most satisfactory planting plan for a fruit- and nut-growing enterprise. There is a trend toward use of dwarfing stocks, growth control chemicals, or closer planting and training, or all of them to get the highest yields and best operation efficiency possible on a unit of ground.

Low-growing crops such as strawberry and pineapple are usually managed in beds containing several rows, or in less formal matted rows. In an acre of strawberries, 200,000 or more plants may occupy the matted rows. A pineapple plantation with two-row beds, having plants one foot (0.3 metre) apart in rows two feet (0.6 metre) apart totals 15,000 to 18,000 plants per acre (37,000 to 44,000 per hectare). With such dense populations, intense competition for light, water, and nutrients causes smaller average fruit size. Nevertheless, the total yield per unit of land is usually greater than it would be with lower plant numbers.

The spacing of grapevines along a trellis row and of trees planted in hedgerows involves the same group of problems. Maximum vineyard production frequently results with vine distances of eight to nine feet (2.4 to 2.7 metres; 600 ± per acre [1,500 per hectare]). The trend for peach trees and spur-type apple strains is hedgerows 14 feet (4.2 metres) apart or closer, in rows 18 to 20 feet (5.4 to 6 metres) apart.

With those species and varieties that require cross-pollination by insects, the planting plan must take those special needs into account. This is a problem with apple, pear, plum, and sweet cherry orchards. At least two varieties that cross-fertilize successfully must be planted in association with each other.

Training and Pruning

Pruning is the removal of parts of a plant to influence growth and fruitfulness. It is an important fruit-growing practice. Primary attention is given to form in the first few years after fruit trees or vines are planted. Form influences strength and longevity of the mature plant as well as efficiency of other fruit-growing practices; pruning for form is called training. As the plant approaches maximum fruitfulness and fills its allotted space, maintenance pruning for various purposes becomes increasingly important.

The grape may be trained following one of two systems: (1) spur system, cutting growth of the previous season (canes) to short spurs, (2) long-cane system, permitting canes to remain relatively long. Whether a spur or long-cane system is followed depends on the flowering habit of the variety. Relatively small trees that respond favourably to severe annual pruning (e.g., the peach and Kadota fig) are usually trained to create an open-centred tree with a scaffold of four or five main branches that originate on a short trunk and branch a number of times to provide fruiting wood. Annual renewal pruning can be reasonably efficient under these circumstances. Larger trees that do not respond favourably to heavy annual pruning are trained best to a system that encourages the main leader branch to grow erect to a height of eight to 10 feet (2.4–3.0 metres), with four or five main lateral branches at intervals on its sides forming the scaffold that carries fruiting wood up and out; this is called a modified leader system. The central leader type of tree, with one main leader up through the centre and many side branches, is common for pear and apple planted in hedgerows, and possibly for other fruits and nuts as the close-planted hedgerow system is more widely adopted.

The principal reasons for maintenance pruning are: (1) to permit efficient spraying and harvesting operations, (2) to maintain satisfactory light exposure for most of the leaves, and (3) to create a satisfactory balance between flowering and leaf surface.

To reduce hand labour costs, larger commercial fruit growers use machine pruning on many types of fruits. Peach, apple, pear, and other fruits usually planted in hedgerows are mowed across the top and sides by machine, and then thinned out as needed by a follow-up crew using pneumatic clippers and hand-powered saws, operating from hydraulically manipulated scaffolds or lifts of various types.

Soil Management, Irrigation and Fertilization

Soil Management

Two soil management practices (1) clean cultivation and chemical weed control or both and (2) permanent sod culture, illustrate contrasting purposes and effects. In clean cultivation or chemical weed control, the surface soil is stirred periodically throughout the year or a herbicide is used to kill vegetation that competes for nutrients, water, and light. Stirring increases the decomposition rate of soil organic matter and thereby releases nitrogen and other nutrients for use by the fruit crop. It may also provide some improvement in water penetration. On the other hand, laying bare the soil surface exposes it to erosion; destruction of organic matter eventually lowers fertility and causes soil structure to change from loose and friable to tight and compacted. Though sod culture minimizes the destructive processes and may permit a modest increase in fertility, the sod itself competes with fruit plants for water and nutrients and may even compete for light. As a result, permanent sod culture is practical only with tree crops that are normally rather low in vegetation, such as apple, pear, sweet cherry, nuts, and mango. Competition from established sod may be detrimental to vigorously growing fruit plants like grape, peach, and raspberry unless adequate fertilizer and water are supplied.

Because each of these soil management systems has advantages and disadvantages, modifying or complementary practices are often used; for example, cover cropping, mulching, and chemical control of vegetation with or without strip sod in the row middles. In fact, the trend is toward mowed sod middles with strip chemical control under the trees and with overhead sprinklers during hot dry weather. Sprinklers not only provide water but tend to cool the plants and give fruit of better market quality without aggravating diseases. Cultivation combined with winter cover cropping has been used widely in grape, peach, cherry, bush fruit, and citrus plantings, as well as with other species. Mulching is the addition of undecomposed plant materials such as straw, hay, or processors' refuse to the soil under the plants. In orchards, mulching materials are most often applied under trees maintained in permanent sod. Strip in-row chemical control of vegetation in commercial fruit plantings has almost taken over as an economical and sound practice.

Irrigation

In semi-arid and arid regions, irrigation is necessary. Probably the maximum demand occurs in date gardens, because they expose a large leaf surface the year around under conditions of high evaporation and practically no rainfall. Irrigation in humid climates is generally being provided increasingly during extended dry periods that occur at one time or another during most growing seasons. For example, large acreages of banana are irrigated on coastal lowlands of the torrid tropics where annual rainfall exceeds 60 inches (1,500 millimetres).

Fertilization

Needs of perennial fruit plants for fertilizers depend on the natural fertility of the soil supporting them and on their individual requirements? Of the essential elements, supplemental nitrogen is almost always needed; potassium supplements may be needed, even in some desert areas. Although strawberry, grape, peach, and a few other fruits have responded favourably to phosphorus, and although its application has been recommended, the phosphorus requirement of woody plants is low and deficiency is rather rare. Calcium deficiency may be more common than realized; lime is often desirable to reduce soil acidity and because of other indirect benefits. Inadequate magnesium in the soil has been noted by workers studying a wide range of fruit species. Of the trace elements, zinc, iron, and boron are most likely to be deficient, but copper, manganese, and molybdenum deficiencies also are being reported for some fruits in some regions. Iron deficiency is difficult to control in orchards where soils have high alkalinity. Granulated fertilizers in modern close-planted commercial orchards are usually broadcast by machine a month or two before growth starts. Additional nitrogen sometimes is applied in heavy crop years to apple, pear, and citrus.

Crop Enhancement

Pollination

The stimulus of pollination, fertilization, and seed formation is needed to get good size, shape, and flavour of most of the fruits. (Banana, pineapple, and some citrus and fig varieties are exceptions.) Transfer of pollen from the anthers (male) to the stigmas (female) is accomplished in nature either by insects or by movement in air. It is common practice to bring beehives into the orchard during bloom. Rainy cold weather during bloom with little or no sunshine can deter activity of the honey bee (the key insect pollinator) and reduce fruit set appreciably. This is one of the main problems not fully solved by fruit researchers. Hand-pollination by daubing collected and preserved pollen onto the stigma (as is done with date palms) sometimes is practiced for other fruits, but this approach is not widespread.

Thinning

Removal of flowers or young fruit (thinning) is done to permit the remaining fruits to grow more rapidly and to prevent development of such a large crop that the plant is unable to flower and set a commercial crop the following year. Thinning is done by hand, mechanically, or chemically. With the date, the pistillate flower cluster is reduced in size at the time of hand-pollination. In the case of certain table grape varieties, some clusters are cut off. With the Thompson seedless grape, a combination of girdling the trunk bark and judicious application of gibberellin (growth regulating) sprays at blossoming gives excellent full bunches.

Young peach fruits are thinned by striking the branches with a padded pole or by shaking the entire tree for a few seconds with a well-padded motor-driven shaker arm grasping the trunk. Hand thinning of young apple and peach fruits once was also a common practice, but because of the expense and difficulty, there has been increasing use of chemical sprays as a substitute. Two kinds of sprays are used: (1) mildly caustic sprays applied during bloom, such as Elgetol in arid regions, or (2) sprays of growth-regulating substances such as 3-CPA (2,3-chlorophenoxy propionamide) applied within a few weeks after bloom in areas with late frosts.

Pest Control and Preservation

In many fruit enterprises, pest control is the most expensive and time-consuming growing practice. Where the concentration of fruit farms in an area warrants it, individual efforts are complemented by legislative measures including quarantine regulations to force removal of pest-laden, unattended orchards. Sometimes the most economical control procedure is biological in nature. There is increased research today to find and multiply parasites that kill fruit crop pests. Such biological methods are necessary as political pressures increase for banning DDT and other chemicals. Selection of varieties that are immune, resistant to attack, or tolerant to specific pests, is a biological control procedure also widely used. Chemical control procedures, however, are relied on most heavily. Air-blast spray or mist-application machinery covering 70 acres (28 hectares) of trees or more in a day is now in common use.

Harvesting And Packing

The proper time to remove a fruit from the tree or plant varies with each fruit and is governed by whether the product will be sold and consumed within hours, or stored for weeks, months, or even a year. Most fruits are harvested as close as possible to the time they are eaten. A few, of which banana and pear are outstanding examples, may be harvested while immature and still ripen satisfactorily. Orange, grapefruit, and some varieties of avocado may be "stored" on the tree for several months after they have attained good quality; this method cuts costs in handling and marketing.

Many fruits, including apple, pear, orange, lemon, and grapefruit, may drop from the tree during the last part of the maturation period. Preharvest drop of these fruits can be delayed by application of dilute sprays of growth-regulating substances like naphthaleneacetic acid (NAA). The chemical spray Alar [N-(dimethylamino) succinamic acid] applied four to six weeks after bloom on apple not only reduces fruit drop at harvest but increases red colour, firmness, and return bloom the next year, in addition to other advantages.

For the fresh market, most tree and bush fruits are still harvested by hand. For processing, drying, and occasionally for fresh market, mechanical motor-driven tree and bush shakers with appropriate catching belts, bins, pallets, and electric lifts reduce harvesting and handling labour. In years to come, machinery may make it possible to machine-harvest most fruits, with no more, and possibly less, damage than with hand picking.

The public has become increasingly particular about the appearance and quality of the product it buys. Hence, store managers and suppliers seek the best grades of fruits and nuts available, and growers make every effort to produce crops with attractive colour and smooth finish. Fruits are packed by government-controlled grades such as Fancy or Extra Fancy within given size limits and are so labelled on the carton or box, together with the source. Most fruits and nuts not meeting this standard of quality are processed or sent through channels using the lower grades and off sizes.

Small packages of plastic foam or wood pulp base holding four to six fruits covered and heat sealed with polyethylene plastic film are popular. These are delivered to stores in corrugated cartons holding a few dozen packages. Citrus, apples, and whole nuts or kernels also are packaged in polyethylene bags and delivered in cartons. Loose fruit may be sold in cell cartons and tray packs consisting of stacked form-fitting pulp trays in a "bushel size" box. Every effort is made to eliminate bruising.

Large truck-pulled containers with individually motor-driven refrigeration units, with or without controlled atmosphere (CO_2-O_2, to retard ripening), are loaded at the fruit source and trucked to their destination or are loaded on ships by derrick for overseas shipment. These sealed containers are also being used increasingly for bananas to reduce labour and handling and to deliver the product in better condition.

Air shipment of "vine- and tree-ripe" fruit (strawberries, figs, sweet cherries, pineapples, avocados) to distances as far as from California to Europe in a day or less is becoming increasingly common with the much larger and faster cargo planes and reduced air-freight prices.

Postharvest Physiology of Fruits

Fruit ripening is a form of senescence and signifies the final stage in fruit development. A fleshy fruit is the enlarged ovary of a flower (avocado) or additional floral parts such as in apple, pear, and pineapple. Usually fertilization, and sometimes pollination alone, stimulates the floral parts causing a rapid cell division that leads to differentiation and the formation of the fruit structure. During this stage fruits consist of small, young cells filled with protoplasm. When the young fruit has been stimulated, presumably by plant hormones that originate from the embryonic seeds, rapid cell expansion takes place. During this stage fruits gain rapidly in size and weight. The cells develop small cavities or spaces in their tissue (become vacuolated) and begin the process of food-stuff accumulation, which lends fruits their compositional diversity. Banana, apple, and date, for example, accumulate mainly carbohydrates. Avocado and olive store fatty materials. Important constituents of most fruits are organic acids such as malic acid, found in apple and pear; citric acid, found in citrus fruits and pineapple; and tartaric acid, found in grapes. Fruits are usually low in protein.

After cell expansion has slowed and become nominal, fruits enter the stage of maturity and undergo preparation for ripening. Some crops, such as pear and avocado, are harvested at the so-called mature-green state and allowed to ripen afterward. Most fruits are at a stage of incipient ripening before they are picked. Ripening is marked by rapid and dramatic changes that give fruits their attractive and edible character. Some of the familiar changes are softening, which results from degradation of cell wall substances; disappearance of a green background, because of chlorophyll degradation (as in pear, apple, and banana); appearance of coloured pigments such as the carotenoids—orange-yellow—and anthocyanins—red (as in orange, mango, and strawberry); a decrease in acidity and increase in the sugar content (orange, apple); and emission of the volatile substances that give many fruits their distinct aroma (as in banana, pear, and apple). In climacteric fruits (e.g., banana, pear, apple), ripening is accompanied by increased respiration. In nonclimacteric fruit (e.g., strawberry, cherry) this phenomenon does not occur.

It is thought that the transition from the mature to the ripe stage is brought about by certain "ripening" enzymes. Protein molecules act as catalysts. The activity of these enzymes leads first to various ripening reactions, and then to gradual deterioration of the fruit tissue.

Because ripening leads to tissue breakdown, fruits are considered a highly perishable commodity. Different fruits have varying degrees of postharvest longevity. While strawberries last only a week to 10 days, for example, apples or lemons can be stored successfully for as long as several months.

Postharvest life of fruits can be extended by refrigeration with or without a modified oxygen–carbon dioxide atmosphere. Most temperate-zone fruits can be held safely at 32 to 41 °F (0 to 5 °C), but many subtropical and tropical fruits, including lemon, avocado, banana, and mango, show signs of injury from being chilled in prolonged cold storage and consequently fail to ripen properly. Bananas do not tolerate temperatures below 53 °F (12 °C), while several avocado varieties can be stored at temperatures as low as 46 °F (8 °C).

Fruit life can be extended further by both refrigeration and controlled atmosphere (CA) storage in which oxygen is kept at about 5 percent and carbon dioxide at 1 to 3 percent, while temperature is held at a level best suited to the particular fruit. So-called CA storage is common today for apples and pears and is being adapted to other fruits. Controlled atmosphere and refrigeration in conjunction with the removal of ethylene gas (which emanates from fruits and speeds ripening) helps slow the ripening process considerably. Golden Delicious apples and some pears are shipped in polyethylene containers in which a desirable, modified atmosphere is created by the respiring fruit.

Drying is a standard practice for stabilizing the market movement of dates, figs, raisin grapes, prunes, and apricots. Canning is of paramount importance to the pineapple, peach, and pear industries (these fruits can be dried as well), and freezing is a means of stabilizing some of the most perishable fruits, including strawberry, raspberry, and blueberry.

Nuts are susceptible to mold, souring, staleness, discoloration, and rancidity. Cured and dried nuts are kept in prolonged cold storage under controlled temperature and humidity levels. Nuts also are stored and sold in vacuum packs of carbon dioxide-enriched atmosphere.

Waste Materials and other Uses

Apple wood is excellent for fireplace use and cherry and certain other fruit woods are used for the finest household furniture. The dried residue from processing apples and citrus is made into feed for conditioning livestock for market, as are waste materials from many processed fruits. Apple pomace (waste material) is spread on the orchard floor with a manure spreader to help in soil conditioning and as a source of minerals.

Nutshells have many uses. Filbert shells are made into plywood, artificial wood, and linoleum; a mixture of shells with powdered coal and lignite makes cinder blocks; shells are used in making poisonous gases and gas masks, and as fuel and mulch. Cashew shell liquid, a skin irritant, is made into resins for varnishes; kills mosquito larvae; can be impregnated in wood as a varnish to preserve against insect attack; is used in automotive brake linings and clutch facings; is used as a laminating agent for paper, cloth, and glass fibres; and is used to treat cement floors and synthetic rubber to retard deterioration. Finely ground black-walnut-shell flour is used in plastic molding powder; as a glue extender; to prevent overheating of drills; to "sand"-blast jet engines; for polishing, burnishing, and deburring metal parts; for cleaning foundry molds; and to spray on tires for better traction. Pecan shells are used in place of gravel in cement walks and driveways; as fuel; as mulch and as a soil conditioner; in livestock bedding; as filler for fertilizers, feeds, etc.; in the manufacture of tanning agents, with charcoal and abrasives in hand soap; as a filler in plastic and veneer wood; and many of the same uses as black walnut shells. Some nutshells are made into beads, marbles, buttons, carving tools, ink, and ornament. The India clearing nut is cut open and rubbed on the inside of earthenware that will contain drinking water; the juice coagulates the water impurities which sink to the bottom. The nuts of the betel palm in the Far East and of the kola tree in West Africa are chewed for their stimulatory effects.

References

- Vegetable: britannica.com, Retrieved 15, May 2020
- Leafy-vegetables: medibiztv.com, Retrieved 06, August 2020
- What-are-root-vegetables, vegetables, side-dishes: southernliving.com, Retrieved 29, April 2020

- Fruit-plant-reproductive-body, science: britannica.com, Retrieved 18, January 2020

- Vegetable-farming-67899: britannica.com, Retrieved 02, July 2020

- Vegetable-farming-from-its-beginnings: blog.agrivi.com, Retrieved 27, February 2020

- Most-profitable-fruit-farming-in-india, economic-importance-of-fruit-farming-than-that-of-agronomic-crops: agrifarming.in, Retrieved 07, June 2020

- Cultivation, fruit-farming: britannica.com, Retrieved 19, March 2020

Chapter 2

Farming of Vegetable and Fruit Crops

There are numerous vegetables which are grown for human consumption such as capsicum, tomato, cucumber, eggplant and pumpkin. A few major fruit crops are guava, banana, apple and strawberry. This chapter discusses in detail these vegetable and fruit crops as well as their farming.

Capsicum

Capsicum is variously called as green pepper, sweet pepper, bell pepper, etc. In shape and pungency it is different from chilli. It is fleshy, blocky, of various shapes, more like a bell and hence named bell pepper. Almost all the varieties of green pepper are very mild in pungency and some of them are non-pungent, and as such they can be used as stuffed vegetable.

Climate

It requires a similar climate like that of chilli and is also susceptible to frost. It prefers milder climate than chilli and 21 to 25 °C is ideal for green pepper. Higher temperatures are detrimental to fruit set. High temperature and low relative humidity at the time of flowering increases the transpiration pull resulting in abscission of buds, flowers and small fruits. Moreover, higher night temperatures are found to be responsible for the higher capsicin (pungency) content in green pepper.

Soil

Although sweet pepper can be grown in almost all types of soils, well drained clay loam soil is considered ideal for its cultivation. It can withstand acidity to a certain extent. Levelled and raised beds have been found more suitable than sunken beds for its cultivation. On sandy loam soils, the crop can be successfully grown provided the manuring is done heavily and the crop is irrigated properly and timely. The most suitable pH range of soil for green pepper is 6 to 6.5.

Varieties

There are a number of varieties of green pepper. The important ones are:

- California Wonder,
- Chinese Giant,
- World Beater,
- Yolo Wonder Bharat,
- Arka Mohlnl,
- Arka Gaurav,
- Arka Basant,
- Early Giant.
- Bullnose, King of North,
- Ruby King, etc.

Planting Requirement

The following procedures are followed to plant capsicum in the field:

Seedling Raising

Seedlings are first raised in the nursery beds and then transplanted in the main fields. Normally, 5-6 seed beds of size (300×60×15 cm) each are sufficient for one hectare cultivation. Seed should be sown in rows at 8 -10 cm apart to get healthy seedlings. The seeds should be dressed with Agrosan. Ceresan, Thiram or Captan @ 2 g per kg seed before sowing to prevent the occurrence of any seed-borne diseases. About 1 -2 kg seeds are required for one hectare cultivation depending on the cultivar. The seeds should be properly covered with a thin layer of soil manure mixture or any other media and irrigated with sprinkler to maintain optimum moisture till the seeds germinate.

Sowing Time

The sweet pepper is generally sown in August for the autumn-winter crop and in November for the spring -summer crop. In the hills of North Bengal sowing of seeds in the months of MarchApril (under cover) and September -October is very successful for getting high yield. Plants sown in September and October take the longest period for development because of poor availability of light in winter.

Land Preparation

For planting the seedlings, The main field is thoroughly prepared by ploughing the land 5-6 time followed by smooth planking. Farmyard manure or compost is added after the first ploughing so that it is thoroughly mixed in the soil during subsequent ploughings. Then the field is brought to a clean and fine tilth.

Transplanting

The seedling having attained 4-5 leaves should be transplanted. The nursery beds should be irrigated before lifting of seedlings. The seedlings are transplanted in rows in the evening or during the cloudy day followed by irrigation. Generally, 50 to 60 days old seedlings are used for transplanting.SpacingThe seedlings are transplanted in rows at a distance of 30 to 60 cm depending upon the area and the variety. Rows spaced at 90 cm and plants spaced at 40 to 45 cm are also fairly common.

Manures and Fertilizers

About 50 to 80 cartloads of farmyard manure, 30 to 55 kg of nitrogen in the form of ammonium sulphate or urea, 50 to 110 kg of phosphorus in the form of super phosphate and 75 to 100 kg of potash per hectare should be given depending upon the fertility status of the soil. The complete dose of farmyard manure should be applied in the soil at the time of first ploughing. Potassium and phosphate fertilizers should be mixed in the plant rows just before transplanting. The nitrogenous fertilizer is given two and half a month after transplanting.

Irrigation

The first irrigation is given just after transplanting and later the field should be irrigated as and when required. Irrigation is essential in arid and semi-arid regions.Weed ControlIntercultural operations are similar to that of chillies. Two weedings 30 and 60 days after transplanting lead to high yield in green pepper. Earthing of plants may also be done after 2 -3 weeks of transplanting. Earthing operation will also help in removing the weeds.

Insects and Diseases

Some of the important insects and diseases which attack capsicum are described below:

Thrips

The adults of these tiny insects are slender yellow, active and pointed at both ends. The females have four extremely slender wings which have long fringe on their posterior margins. The male is similar to female except that it is smaller and lighter in colour. The minute insects lacerate the plant tissues and suck the sap from the leaves forming white blotches and curly leaves with stunted plant look. Consequently yield is reduced considerably.

- It can be controlled by the spraying Malathion (Cythion 50 EC @ 1.5 m1 per litre) or Dimethoate (Rogar 30 EC @ 1 ml per litre of water). It may also be controlled by the spraying of 0.25% Nicotine sulphate.AphidsAphids sometimes becomes serious on capsicum. They suck the cell sap from the leaves and petioles and cause considerable loss.

- Complete control of aphids can be obtained by the application of Dimeton methyl (0.05 to 0.02%) or Monocrotophos (0.05 to 0.01%).

- Mites of different genera have been found feeding on leaves of chilli and capsicum. These tiny spider like creatures may be found in large numbers on the underside of leaves, covered with fine webs. Both nymph and adults suck the cell sap and devitalise the plants.

Control

It is reported that spraying of Phosaione (Zolone) 35 EC at 3 ml per litre can controi mites. -Spraying Dimethoate (Rogar @ 1 ml per litre) or Dicophol (Kelthane @ 1.5 mi per litre of water) is very effective against mites.

Harvesting and Yield large sweet peppers usually are picked while they are still green in colour but fully grown when sold in the market. Some exotic varieties such as Pimiento and Paprika are harvested when fruits are GMk red ripe. However, the most favourable time for harvesting in

Pimiento for seed production is between 50 -60 day old stage or when the fruit attain the bright to deep red colour. Sweet peppers are picked with an upward twist which leaves a piece of stem attached with the fruits. Young immature peppers are rather soft and yield readily to mild pressure of the fingers. Green fruits ready for harvest are relatively firm and crisp. The yield of capsicum varies depending upon variety and the method of cultivation. If proper care is taken during its growth, it may yield 10 to 12 tonnes of quality fruits per hectare.Ripening and StorageThe bell type peppers are usually harvested and sold when they are of suitable market size and are still green. There is but a limited demand for the mature red specimens. Now-a-days different chemicals are used to accelerate ripening as well as inducing the fruit colouration. Spraying with Ethephon at 200 to 3200 ppm has been found effective to accelerate fruit colour development, fruit and leaf drop and leaf yellowing. Some varieties are harvested when they are completely red. Green peppers can be kept in good condition for at least 40 days at 00 C and at 95 to 98% relative humidity. The shrinkage of fruits stored under these conditions is only 4% in 40 days.Seed ProductionIn seed to seed method of seed production, transplanting is done for commercial seed production. Optimum spacing may be followed to obtain high quality seeds. Capsicum is a cross pollinated crop, so it crosses easily with chillies and thus deteriorates fast in quality, if proper isolation distance is not maintained during seed production stages.

The isolation distance between two cultivars of capsicum should be 200 meters for foundation seed and about 100 meters for certified seed production. Off-type plants are removed as soon as they are observed. The small leaved plants can be detected from the large leaved plants and should be rogued as per the requirement. The number of rouging should be 3 -4 depending upon the purity of the seed desired. Field inspection of seed crops should be done at least twice or thrice. The fruit should be picked when red ripe and cut and crushed or macerated by machine. Seed is to be washed thor-oughly to make it free from pulp and skin. After washing, the seeds should be dried immediately in the sun. Picking of pods may be done according to climatic conditions. In case, there is no danger of rains at the maturity time, the pods may be picked in one lot but where there is some danger of rains picking may be done is 2 -3 instalments.

Tomato

Tomato (*Lycopersicon esculentum*) is an annual or short lived perennial pubescent herb and greyish green curled uneven pinnate leaves. The flowers are off white bearing fruits which are red or yellow in colour. It is a self-pollinated crop.

Soil and Climate

Soil

Tomato can be grown on a wide range of soils from sandy to heavy clay. However, well-drained, sandy or red loam soils rich in organic matter with a pH range of 6.0-7.0 are considered as ideal.

Climate

Tomato is a warm season crop. The best fruit colour and quality is obtained at a temperature range of 21-24 °C. Temperatures above 32 °C adversely affects the fruit set and development. The plants cannot withstand frost and high humidity. It requires a low to medium rainfall. Bright sunshine at the time of fruit set helps to develop dark red coloured fruits. Temperature below 10 °C adversely affects plant tissues thereby slowing down physiological activities.

Varieties

- Released by IARI : Pusa Rohini, Pusa Sadabahar, Pusa Hybrid 8, Pusa Hybrid 4, Pusa Uphar, Pusa Hybrid 2, Sioux.

- Released by IIHR : Arka Vikas, Arka Saurabh, Arka Meghali, Arka Ahuti, Arka Ashish, Arka Abha, Arka Alok, Arka Vishal, Arka Vardan, Arka Shreshta, Arka Abhijit.

- Released by PAU : Pb. Kesari, Punjab Chhuhara, S-12, Sel-152, PAU-2372.

- Released by GBPUAT, Pantnagar : Pant T-10, AC-238, Pant T-3.

- Others : H-24, H-86, Pusa Early Dwarf, CO-3, CO-1, BT-12.

Propagation

Nursery Bed Preparation

Tomato seeds are sown on nursery beds to raise seedlings for transplanting in the field. Raised beds of size 3 × 0.6 m and 10-15 cm in height are prepared. About 70 cm distance is kept between two beds to carry out operations of watering, weeding, etc. The surface of beds should be smooth and well levelled. Add sieved FYM and fine sand on the seedbed. Raised beds are necessary to avoid problem of water logging in heavy soils. In sandy soils, however, sowing can be taken up in flat beds. To avoid mortality of seedlings due to damping off, drench the seed bed first with water and then with Bavistin (15-20 g/10 litres of water).

Season of Planting

Seeds are sown in June July for autumn winter crop and for spring summer crop seeds are sown in November. In the hills seed is sown in March April.

Raising of Seedlings

About 250-300 g of seed are sufficient for raising seedlings for one hectare of land. Prior to sowing seeds are treated with fungal culture of *Trichoderma viride* (4 g/ kg of seed) or Thiram (2g/kg of seed) to avoid damage from damping-off disease. Sowing should be done thinly in lines spaced at 10-15 cm distance. Seeds are sown at a depth of 2-3 cm and covered with a fine layer of soil followed by light watering by water can. The beds should then be covered with dry straw or grass or sugarcane leaves to maintain required temperature and moisture. The watering should be done by water can as per the need till germination is completed. The cover of dry straw or grass is removed immediately after germination is complete. During the last week in nursery, the seedlings may be hardened by slightly withholding water. The seedlings with 5-6 true leaves are ready for transplanting within 4 of sowing.

Planting

Land Preparation

The field is ploughed to fine tilth by giving four to five ploughing with a sufficient interval between two ploughing. Planking should be done for proper levelling. Furrows are then opened at the recommended spacing. Well-decomposed FYM (25 t/ha) is thoroughly incorporated at the time of land preparation.

Spacing

Spacing depends upon the type of variety grown and the season of planting. Normally the seedlings are transplanted at a spacing of 75-90 x 45-60 cm.

Method of Planting

Seedlings are transplanted in furrows in light soils and on side of the ridges in case of heavy soils. A pre-soaking irrigation is given 3-4 days prior to transplanting. Before planting seedlings should be dipped in a solution prepared by Nuvacron (15ml) and Dithane M - 45 (25g) in 10 litres of water for 5-6 minutes. Transplanting should preferably be done in the evening.

Inter Cultivation

Weed Control

The field should be kept weed-free, especially in the initial stage of plant growth, as weeds compete with the crop and reduce the yield drastically. Frequent shallow cultivation should be done at regular interval so as to keep the field free from weeds and to facilitate soil aeration and proper root development. Deep cultivation is injurious because of the damage of roots and exposure of moist soil to the surface. Two-three hoeing and the earthing up are required to keep the crop free of weeds. Pre- emergence application of Basalin (1kg a.i./ha) or Pendimethalin (1kg a.i./ha), coupled with one hand weeding 45 days after transplanting is effective for control of weeds. Plastic mulching (black or transparent) can be used to control the weeds. Weeds can be controlled successfully by mulching plus use of herbicides such as Pendimethalin (0.75 kg a.i./ha) or Oxyfluorfen (0.12 kg a.i./ha).

Crop Rotation

Tomato should not be grown successively on the same field and a break of at least one year is required between planting of tomatoes or other Solanacesous crops (eg. Chillies, Brinjals, Capsicum, Potato, Tobacco, etc.), cucurbits and many other vegetables. The crops, which can be grown after tomatoes, are as follows- Cereals (eg. Rice, Corn Sorghum, Wheat, Millets, etc.) or Cruciferons crops (eg. Cabbage, Cauliflower, Kohlrabi etc) or Radish, Watermelon, Onion, Garlic, Groundnut, Cotton, Safflower, Sunflower, Sesame, Sugar beet and Marigold.

Intercropping

Tomato is well fitted in different cropping systems of cereals, grains, pulses and oilseeds. Cropping systems rice-tomato, rice-maize, okra-potato-tomato, tomato-onion are popular. Spinach or radish can also be grown as inter-crop in tomato successfully.

Staking

Due to the tall habit and heaving bearing nature of the hybrids staking is essential. Staking facilitates intercultural operations and helps in maintaining the quality of the fruits. It is done 2-3 weeks after transplanting. Staking can be done either by wooden stakes or laying overhead wires to which individual plant is tied. In case of indeterminate types, tow or three wires are stretched parallel to each other along the row and plants are tied to these wires.

Irrigation

Tomato is very sensitive to water application. Heavy irrigation provided after a long spell of drought causes cracking of the fruits. Hence it should be avoided. Light irrigation should be given 3-4 days after transplanting. Irrigation intervals should be according to soil type and rainfall, irrigation should be given 7-8 days interval during kharif, during rabi 10-12 days and 5-6 days during summer.

Flowering and fruit development are the critical stages of tomato therefore; water stress should not be given during this period.

Manuring & Fertilization

The fertilizer dose depends upon the fertility of soil and amount of organic manure applied to the crop. For a good yield, 15-20 tonnes of well-decomposed FYM is incorporated into the soil. Generally, application of 120 kg N, 80 kg P_2O_5 and 50 kg K_2O per hectare is recommended for getting optimum yield. Half dose of N and full dose of P and K is given at the time of planting. The balance half of N is given as top dressing 30 days after transplanting.

For hybrid varieties, the recommended dose per hectare is 180 kg N, 100 kg P_2O_5 and 60 kg K_2O. 60 kg N and half of P & K are given at the time of transplanting. Remaining quantities of P & K and 60 kg N is top dressed 30 after transplanting. A third dose of 60 kg N is applied 50 days after transplanting.

Growth Regulators

Effect of growth regulators in tomato crop is as follows:

Plant-growth regulators	Concentration (mg/litre)	Method of application	Attributes affected
Gibberellic acid (GA)	10-20 40-100	Foliar spray Seed treatment.	Higher yield at low temperature Seed germination.
Ethephon	100-500 1,000	Foliar spray Pre-harvest spray.	Flowering, fruiting and yield Fruit ripening.
PCPA	50-100	Foliar spray at low flowering.	Tomato fruit set at high temperatures.

IPM Practices for Tomato Pests

The IPM package given below will take care of fruit borer, leaf miner, mite and insect vector.

Nursery

- Raise Marigold (Tall African variety golden age bearing yellow and orange flowers) nursery 15-20 days before tomato nursery.

- One week after germination of seeds, spray the seedlings with (imidacloprid 200 SL @ 0.3 ml/l or thiomethoxam 25 WP @ 0.3 g/l).

Before Transplanting

- Apply Neem cake 250 kg/ha ridges at the time of preparing land.

- Dip the roots of seedlings (do not dip the foliage as it may cause burning of leaves) with imidacloprid 200 SL @ 0.3 ml/l or thiomethoxam 25 WP @ 0.3 g/l for 5 minutes.

Main Field

- Transplant 20-25 day old tomato and 45-50 day old marigold simultaneously in a pattern of one row of marigold for every 16 rows of tomato. However, the first and last row of the plots should be of marigold. Simultaneous flowering of both the crops ensures attraction of fruit borers to marigold flowers.

- Fifteen days after planting spray imidacloprid 200 SL @ 0.4ml/l or thiomethoxam 25 WP @ 0.3g/l for leaf curl vector (whitefly) control.

- Apply neem cake @ 250 kg/ha to ridges at 20-25 DAP (at flowering) to reduce nematode, fruit borer and leaf miner incidence.

- Spray Ha NPV (@250 LE/ha) with 1% jaggery as sunscreen at 28, 35 and 42 DAP in the evening.

- Spray marigold flowers with *Ha*NPV or destroy fruit borer larvae in them.

- As an alternative to *Ha*NPV spray, the egg parasitoids, *Trichogramma chilonis, T. braziliensis and T. pretiosum* @ 2.5 lakhs/ha can be released (five releases @ 50,000/ha/ release). The first release has to be done at the flower initiation of the crop.

- If red spider mite incidence is noticed, spray Neem soap 1 % or neem oil 1% or any synthetic acaricide like dicofol 18.5 EC (1.5 ml/l), or Ethion 50 EC (1.5 ml/l) or sulphur 80 WP (3 g/l) etc. Spray lower surface of the leaves.

- Mechanical collection and destruction of bored fruits at periodic interval (3-4 times after fruit set) to minimize fruit borer incidence to the minimum.

- Destroy leaf curl and other virus affected plants as soon as the symptoms appear in a few plants to minimize their spread.

Harvesting

Depending on the variety, fruits become ready for first picking in about 60-70 days after transplanting. The stage of harvesting depends upon the purpose to which the fruits are to be used. The different stages of harvesting are as follows:

- Dark green colour: Dark green colour is changed and a reddish pink shade is observed on fruit. Fruits to be shipped are harvested at this stage. Such fruits are then sprayed with ethylene 48 hours prior to shipping. Immature green tomatoes will ripen poorly and be of low quality. A simple way to determine maturity is to slice the tomato with a sharp knife. If seeds are cut, the fruit is too immature for harvest and will not ripen properly.

- Breaker stage: Dim pink colour observed on ¼ part of the fruit. Fruits are harvested at this stage to ensure the best quality. Such fruit are less prone to damage during shipment often fetch a higher price than less mature tomatoes.

- Pink stage: Pink colour observed on ¾ part of the fruit.

- Reddish pink: Fruits are stiff and nearly whole fruit turns reddish pink. Fruits for local sale are harvested at this stage.

- Fully ripped: Fruits are fully riped and soft having dark red colour. Such fruits are used for processing.

Fruits are normally harvested early in the morning or evening. The fruits are harvested by twisting motion of hand to separate fruits from the stem. Harvested fruits should be kept only in basket or crates and keep it in shade. Since all the fruits do not mature at the same time, they are harvested at an interval of 4 days. Generally there will be 7-11 harvests in a crop life span.

Yield

The yield per hectare varies greatly according to variety and season. On an average, the yield varies from 20-25 t/ha. Hybrid varieties may yield up to 50-60 t/ha.

Cucumber

Cucumber is a summer vegetable, with elongate shape and 15cm long. Its skin is of a green colour, turning into yellow in maturation.

Growing Cucumbers Requirements and Habits of the Plant

The cucumber responds like a semitropical plant. It grows best under conditions of high temperature, humidity, and light intensity and with an uninterrupted supply of water and nutrients. Under favourable and stable environmental and nutritional conditions and when pests are under control, the plants grow rapidly and produce heavily. The main stem laterals, and tendrils grow fast. They need frequent pruning to a single stem and training along vertical wires to maintain an optimal canopy that intercepts maximum light and allows sufficient air movement.

Under optimal conditions, more fruit may initially develop from the axil of 4 each leaf than can later be supported to full size, so fruit may need thinning. Plants allowed to bear too much fruit become exhausted, abort fruit, and fluctuate widely in productivity over time. Excessive plant visor is indicated by: rapid growth, thick and brittle stems large leaves, long tendrils, deep green foliage, profusion of fruit, and large, deep yellow flowers. On the other hand, cucumbers are very sensitive to unfavorable conditions, and the slightest stress affects their growth and productivity.

Because fruit develops only in newly produced leaf axils, major pruning may be needed to stimulate growth. The removal of entire weakened laterals is more effective than snipping back their tips.

Important Growing Cucumbers Parameters

Temperature, General and with Special Reference to Greenhouses

Air temperature is the main environmental component influencing vegetative growth, flower initiation, fruit growth, and fruit quality. Growth rate of the crop depends on the average 24-h temperature the higher the average air temperature the faster the growth. The larger the variation in day night air temperature, the taller the plant and the smaller the leaf size. Although maximum growth occurs at a day and night temperature of about 28 °C, maximum fruit production is achieved with a night temperature of 19-20 °C and a day temperature of 20-22 °C. The recommended temperatures in table are therefore a compromise designed for sustained, high fruit productivity combined with moderate crop growth throughout the growing season.

During warm weather (i.e., late spring and early fall), reduce air temperature settings, especially during the night, by up to 2 °C to encourage vegetative growth when it is retarded by heavy fruit load. This regime saves energy because a 24-h average can be ensured by the prevailing high temperatures and favorable light conditions.

Table: Recommended air temperatures for cucumber cropping.

	Low light (°C)	High light (°C)	With carbon dioxide (°C)
Night minimum*	19	20	20
Day minimum	20	21	22
Ventilation	26	26	28

*A minimum root temperature of 19 °C is required, but 22-23 °C is preferable.

The optimum daily average air temperature is 15-24 °C (65-75 °F). Optimum temperatures for growth are at night, about 18 °C, and during the day, about 28°C accompanied by high light intensity.

To ensure satisfactory stand establishment, soil temperatures should be at least 15 °C (60 °F). The higher the soil temperature, the more rapidly seedlings emerge and the less vulnerable they are to insects and damping-off diseases.

At 15 °C (60 °F), 9 to 16 days are required for seedlings to emerge. At 21 °C (70 °F), only 5 to 6 days are required. Even after emergence, cucumbers remain sensitive to cold temperatures. In cold areas, seeds should always be planted late enough to avoid frosts. Exposure to cool conditions will slow growth even if temperatures remain above freezing. Slow-growing seedlings are vulnerable to flea beetles (whose chewing significantly reduces leaf area of young plants). Too high temperatures during flowering decrease pollen viability. Cool and cloudy growing season may cause bitter fruit.

Light with Special Reference to Greenhouses

Plant growth depends on light. Plant matter is produced by the process of photosynthesis, which takes place only when light is absorbed by the chlorophyll (green pigment) in the green parts of the plant, mostly the leaves. However, do not underestimate the photosynthetic productivity of the cucumber fruit, which, because of its size and color, is a special case. In the process of photosynthesis, the energy of light fixes atmospheric carbon dioxide and water in the plant to produce such carbohydrates as sugars and starch.

Generally, the rate of photosynthesis relates to light intensity, but not proportionally. The importance of light becomes obvious in the winter, when it is in short supply. In the short, dull days of late fall, winter, and early spring, the low daily levels of radiant energy result in low levels of carbohydrate production.

Not only do the poor light conditions limit photosynthetic productivity, but also the limited carbohydrates produced during the day are largely expended by the respiring plant during the long night. The low supply of carbohydrates available in the plant during the winter seriously limits productivity, as evidenced by the profusion of aborted fruit. A fully grown crop benefits from any increase in natural light intensity, provided that the plants have sufficient water, nutrients, and carbon dioxide and that air temperature is not too high.

Relative Humidity with Special Reference to Greenhouses

High relative humidity (RH) generally favors growth. However, reasonable growth can be achieved at medium or even low relative humidity. The crop can adjust to and withstand relative humidity from low to very high but reacts very sensitively to drastic and frequent variation in relative humidity. Its sensitivity to such variation is greatest when the crop is developed under conditions of high relative humidity.

Other disadvantages of cropping under conditions of high relative humidity include the increased risk of water condensing on the plants and the development of serious diseases. The resultant low transpiration rates are blamed for inadequate absorption and transport of certain nutrients, especially calcium to the leaf margins and fruit. At low relative humidity, irrigation becomes critical, because large quantities of water must be added to the growth medium without constantly flooding the roots and depriving them of oxygen. Furthermore, low relative humidity favors the growth of powdery mildew and spider mites, which alone can justify installing and operating misting devices.

Carbon Dioxide in Greenhouses

In relatively high temperatures and light intensity, supplemental carbon dioxide applied at a concentration up to 400 ppm has proved economically useful. Regions with a moderate maritime climate, can more likely benefit from carbon dioxide applied in the summer only. But in regions with a continental climate, the need to ventilate the greenhouse actively throughout the hot summer renders the practice less economical. Apply carbon dioxide during the day or any part of the night when artificial light is supplied. It is economically feasible and highly advisable to use liquid carbon dioxide (carbon dioxide gas liquefied under pressure) because of its guaranteed purity and amenity to accurate concentration control. Liquid carbon dioxide is also preferred because burning natural gas or propane to generate carbon dioxide increases the risk of plant injury from gaseous pollutants, e.g. ethylene.

Seed Germination

Seeds germinate and emerge in three days under optimum conditions. During this time seed coat remains tight. Once cotyledons emerge, roots develop quickly. Sunlight delivers photosynthates to true leaves and root system.

During the 1st week cotyledons integrity is very important, and if damaged, plants will set back. Seedlings may recover but they will be weak and susceptible to stresses. For proper germination, soil temperature must be above 15 °C (60 °F). If the soil is too cold and wet poor seedling emergence will take place.

Planting

Cucumbers growth season is relatively short, lasting 55-60 days for field-grown varieties, and over 70 days for greenhouse varieties.

Planting Dates

Cucumbers are almost always direct seeded. Like most cucurbits, they do not transplant well and transplant costs would be hard to recover. Planting depth is 2.5-4 cm (1-1.5 inches). Too deep delays emergence. Pickling cucumbers have to be very precise on planting dates so that harvest will coincide with processor needs.

For early crop, container-grown transplants are planted when daily mean soil temperatures have reached 15 °C (60 °F) but most cucumbers are direct seeded. Early plantings should be protected from winds with hot caps or row covers. Growing on plastic mulch can also enhance earliness.

Spacing

Planting spacing depend on the growth method, variety and harvesting method. Close spacing increases yields, provides more uniform maturity and reduces weed problems. It also results in shorter fruit with a lighter color. On the other hand, high plant population requires more seeds and slightly higher fertilizer rates.

Table: Spacing and seeding parameters for open field cultivation.

Hand- harvested							
Consumption method	Plant density (plants/ha)	Distance between rows		Distance within row		Seeds mass	
		cm	Feet	cm	Inch	kg/ha	lbs/acre
Fresh (slicers)	40-50,000	90-120	3-4	23-30	9-12	1.7	1.5
Pickles (processing)	60-75,000	90-120	3-4	15-20	6-8	2.0-5.5	2-5
Machine- harvested							
Pickles (processing)	50-75,000	60-70	24-28	5-10	2-4	1.8-5.5	1.8-5

Greenhouse plant spacing should provide: 1-2.5/m2 or more per plant, depending on pruning and training system. Recommended density is 33,000 - 60,000 plants/ha.

cucumber greenhouse with trellised plants.

Soils

Cucumbers prefer light textured soils that are well drained, high in organic matter and have a pH of 6 - 6.8. Adapted to a wide-range of soils, but will produce early in sandy soils. Cucumbers are fairly tolerant to acid soils (down to pH 5.5).

Greenhouse cucumbers generally grow quite well in a wide range of soil pH (5.5-7.5), but a pH of 6.0-6.5 for mineral soils and a pH of 5.0-5.5 for organic soils are generally accepted as optimum. When the pH is too low, add ground calcitic limestone, or an equal amount of dolomitic limestone when the magnesium level in the soil is low, to raise it to a desirable level.

Usually the pH in most greenhouse mineral soils is above the optimum pH range (6.0-6.5). A simple, though temporary, solution to a high pH problem is to add peat, without neutralizing its acidity with limestone. Peat also helps to maintain a good soil structure, but it must be added yearly to make up for loss through decomposition.

Soil Preparation Before Field Planting

- Soil fumigation aids in the control of weeds and soil-borne diseases. Fumigation alone may not provide satisfactory weed control under clear plastic.

- Black plastic mulching before field planting conserves moisture, increases soil temperature, and increases early and total yield. Plastic should be placed immediately over the fumigated soil. The soil must be moist when laying the plastic. Black plastic can be used without a herbicide.

- Plastic and fumigant should be applied on well prepared planting beds, 2-4 weeks before field planting.

- Fertilizer must be applied during bed preparation. At least 50% of the nitrogen (N) should be in the nitrate (NO_3) form.

- Herbicides recommended for use on cucumbers may not provide satisfactory weed control when used under clear plastic mulch on non-fumigated soil.

- Foil and other reflective mulches can be used to repel aphids that transmit viruses in fall planted cucumbers.

- Direct seeding through the mulch is recommended for maximum virus protection. Fumigation will be necessary when there is a history of soilborne diseases in the field. Growers should consider drip irrigation with plastic mulch.

Trellising

Cucumber vines can be trained on trellises to save space and improve yield and fruit quality. But the high cost of trellising makes commercial production by this method uneconomical in most cases. Greenhouse cucumbers must be trellised, because the long fruit bend if they rest on the ground.

The major advantages of trellising a cucumber crop include:

- Harvesting efficiency.

- Pest management efficiency.

- Straighter fruits.

- Uniform fruit color.

- Reduction of fruit loss due to soil diseases.

- Increased yield due to closer rows.

- Reduced rate of crooked fruits makes trellising absolutely necessary for oriental slicing cucumbers.

Disadvantages include:

- Extra cost of trellising materials.

- Labor to erect the system, dismantle it and training the vines.

Different methods of trellising cucumber plants.

Greenhouse Cucumbers

Flowering Habit and Fruit Set

All European greenhouse varieties produce fruit without pollination. They are gynoecious in flowering habit, and fruits develop without the need for pollination.

Cultural Practices

Greenhouse cucumbers grow rapidly under optimum environmental conditions, and fruit production begins 60 - 70 days after seeding. For good production, a temperature range of 24 – 27 Co (75 - 80 Fo) during the day is desirable. While peak daytime temperatures of 29 - 25 Co (85-90 Fo) are tolerable, prolonged periods of high temperatures may adversely affect fruit quality. Night temperatures no lower than 18 Co (65 Fo) will allow a rapid growth rate and earliest fruit production.

Planting and Plant Growing Cucumbers

Plantings of greenhouse cucumbers are ordinarily started from transplants, but direct-seeding in greenhouse beds may also be practical for late summer or early fall plantings, when the time from seeding to fruiting may not be as critical and prevailing temperatures are warm enough for good seed germination without having to heat the greenhouse.

Transplanting makes more efficient use of greenhouse space, because seed germination and early growth of plants can be confined to a smaller nursery area. Disadvantages to transplanting are the costs of containers and the labor costs of transplanting.

Spacing, Training and Pruning

The decision on the number of plants to be grown in a given area of greenhouse should be based upon expected light conditions during growth of the crop and also upon the method of pruning the plants. With good sunlight, each plant is allotted about 0.5 m² (5 feet²) space. Nearly twice that much space may be needed with low light to avoid leaf overlapping and shading by adjacent plants.

Spacing between rows and plants within the row can vary with grower preference. Rows are often spaced 1.2 – 1.5 m (4 - 5 feet) apart, with plants 30 – 45 cm (12 - 18 inches) apart in the row.

Most growers prune their plants by the umbrella system. In this system, all lateral branches are removed as they develop until the plant reaches the overhead support wire. Fruits should not be allowed to develop on the lower 75 cm (30 inches) of the main stem to encourage the plant's rapid vegetative development. Main-stem fruits above that point are allowed to develop at the base of each leaf. More than one fruit may begin to develop at each node. Some growers thin these to a single vigorous fruit, but it may be more practical to leave all young fruits attached, because it has been observed that more than one may mature.

Fruit Thinning

Overbearing can sometimes be a problem. To prevent the plants becoming exhausted and to improve fruit size, control the number of the fruit per plant through selective fruit thinning. This technique is powerful, so use it with great caution. The optimum number of fruits per plant varies with the cultivar and, even more, with the growing conditions. Although, limiting the number of fruits per plant invariably results in premium-priced large fruit, growers risk underestimating the crop's potential or failing to forecast good weather. They may decide to remove too many fruits and thus unnecessarily limit production. Fruit thinning is undoubtedly most useful in the hands of experienced growers who can use it to maximize their financial returns. Fruit to be pruned must be removed as soon as it can be handled, before it grows too large.

Irrigation

Maximum yields and fruit quality will be realized only if the plants receive adequate and timely moisture. Cucumber plants have shallow roots and require ample soil moisture at all stages of growth. When fruit begins setting and maturing, adequate moisture becomes especially critical.

The objective of watering is to maintain a fully adequate supply of water to the plant roots without soaking the soil to the extent that air cannot get to the roots. Do not wait until the plants start to wilt. A good practice is to dig into the soil and judge how much water remains before starting the next irrigation. Regular watering on the same day of the week is unwise. The needs of the plants change daily and seasonally. Water young plants planted in the greenhouse in January or February only once every 5-10 days and then only enough to wet 15-20 cm of the soil. Similar plants growing in June may need 5-10 fold as much water. Let the soil texture and structure determine how much water to add at each application. By examining the soil before watering and several hours thereafter, you can assess the effectiveness of the water application. Depending upon soil type and growth conditions, approximately 25–50 mm of water per week is needed to obtain high quality cucumbers. An irregular water supply, particularly during blossoming and fruit development, can detrimentally affect fruit quality and result in increased nubins or hooked fruit.

The pH of the water is also critical and may need to be adjusted. The target pH of the nutrient solution supplied to the plants should be between 5.5 and 6.0. Nitric, sulfuric, or phosphoric acid are recommended for reducing the pH if it is above 7. If the source water is alkaline due to high bicarbonate concentrations, the pH should be adjusted before the fertilizer salts are added to prevent precipitation. If it is necessary to raise the pH, potassium hydroxide is usually used. Irrigation timing should not interfere with pollination and should allow surfaces to dry before nightfall.

The drip irrigation system is better than most other conventional soil irrigation systems, because it can be used to control crop growth by regulating the supply of water and nutrients. The drip system also allows reduced relative humidity (RH) in the greenhouse, because not all the soil is irrigated and because the system is compatible with the use of white polyethylene film as a light reflecting mulch. In countries suffering from lack of good irrigation water the drip irrigation method has an important feature expressed by high water efficiency leading to very remarkable water savings.

Resources, including irrigation water and energy, are thus used more efficiently with this system. In most cases, use common in-line drippers with a standard flow of 2L/h and one dripper per plant. However, because of the shallowness and extensiveness of the cucumber root system, consider a 4-L/h dripper, which usually results in more lateral movement of the irrigation water. Even better, provide two drippers (2 L/h each) per plant.

Micro-sprinklers, or misters, have also been tried at ground level along the row of plants, with favorable results on root growth, plant vigor, and productivity. However, such irrigation systems, if not properly managed, can easily lead to overwatering and then to disease outbreaks caused by excessive humidity and plant stem wetness. Another alternative is to use lay-flat polyethylene tubes (about 5 cm ID), with small holes spaced 10 cm apart. This system usually applies water to a much larger area than the drip system but does not raise the RH as much as a micro-sprinkler or micro-mister system. Although fairly inexpensive, the lay-flat tube irrigation system has a short lifespan, which requires its frequent replacement. It is not a good choice for large greenhouses, because the water delivery rates vary significantly along the length of the line (i.e., not pressure compensating). Irrigate the crop up to four times a day, and use the irrigation system to apply fertilizer to the crop.

Fertilization

Greenhouse cucumbers grow quickly and should never be allowed to suffer from lack of water or nutrients. Advanced cucumber cultivation must supply the crop with optimal rates of nutrients throughout the growth cycle in the most efficient manner possible, and without degrading soil and water resources. The nutrient uptake rate by greenhouse cucumbers is very high. One study indicates that cucumbers may require in the range of 28 kg/ha (25 lbs/acre) of nitrogen, 5 kg/ha (5 lbs/acre) of phosphorus, and 40 kg/ha (35 lbs/acre) of potassium per week during peak fruit production. Fertilizer management practices need to assure that plant requirements are satisfied to achieve good yields of high-quality fruit. An important method that can assure that plant requirements are really satisfied is leaf analysis.

Leaf Analysis

Visual diagnosis of disorders can be confused by symptoms induced by non-nutritional factors such as disease, pests and chemicals. Leaf analysis can be used to confirm a visual diagnosis. This involves chemical testing of leaves to establish whether specific nutrients are present in plant tissue at normal concentrations. This technique can be used to check the suitability of a fertilizer program and to anticipate the need for nutrient supplements. A disadvantage of leaf analysis is that it is slow, as most laboratories will take at least a week to process samples and report the results back to the grower. In many cases, laboratories do not interpret results or recommend how to remedy the situation.

Nutrigation (fertigation) (the application of fertilizer through the irrigation system) is a popular and efficient method of fertilizing field and greenhouse vegetables. Fertilizers are either dissolved in a large holding tank and the solution pumped to the crop, or they are mixed in concentrated stock solutions, which are then incorporated, using fertilizer injectors, into the irrigation water.

Several makes and models of fertilizer injectors are available at varying costs and offer varying degrees of fertigation control. A sophisticated fertilizer injection system capable of meeting the nutrient requirements of a series of crops automatically from the same set of stock solutions was developed may use an IBM-compatible computer to activate a series of dosimetric pumps at varying frequencies for the pre-programmed application of a desired concentration of individual nutrients. It also automatically adjusts the supply of water and nutrients to the crops in accordance with crop and environmental conditions. Introducing drip irrigation and fertigation has made it necessary to consider the fertilizer needs of a crop in terms of the concentration of fertilizer (and therefore the concentration of nutrients) in the irrigation water rather than on the basis of the cropped area.

Furthermore, the recommendations regarding the nutrient content of the fertigation solutions of drip irrigated crops are based mainly on the physiological responses and commercial productivity of the crops. Earlier recommendations for fertilizer application to traditionally grown crops in soil were based on estimates of nutrient removal by the crop.

Base fertilizers are not normally applied when drip irrigation is used, but peat and lime, may be still needed pre-planting to improve soil structure and adjust soil pH. Two general examples of fertigation regimes are given, based on leaf analyses done during growth season, (dealing with the entire aspect of cucumber mineral nutrition.

Table: Nutrigation schedule and rates for cucumber with N:K ratio = 1:2

Days after planting	Daily nitrogen	Daily potash	Cumulative			
			nitrogen	potash	nitrogen	potash
			kg/ha		Lbs/acre	
(preplant)			28	50	25.0	45.0
0 - 14	0.9	1.8	42	84	37.6	75.2
15 - 63	1.5	3.0	124	220	110.3	196.6
64 - 77	0.7	1.4	135	243	120.1	216.6

Table: Nutrigation schedule and rates for cucumber with N:K ratio = 1:1 *

Days after planting	Daily nitrogen	Daily potash	Cumulative			
			nitrogen	potash	nitrogen	potash
			kg/ha		Lbs/acre	
(preplant)			27	27	24.0	24.0
0 - 7	1.0	1.0	35	35	31.0	31.0
8 - 21	1.5	1.5	58	58	52.0	52.0
22 - 63	2.0	2.0	152	152	136.0	136.0
64 - 70	1.5	1.5	168	168	150.0	150.0

*Based on leaf analysis.

Nutrigation is a well-established method to increase cucumbers productivity.

Crop Rotation

It is recommended not to rotate to crops in the same family during a 3-year cycle for most diseases, but 5 years must be maintained after Fusarium wilt incidence.

Cucumbers Phytophthora blight is the same as buckeye rot on tomatoes. All curucbits get it, although there are differences in susceptibility. Therefore, cucumbers should be rotated with crops other than peppers, eggplants, tomatoes, and other cucurbits.

Specific Sensitivities

Cucumbers and other cucurbits have a number of specific sensitivities that may jeopardize yields and fruit quality. Among these sensitivities we can mention general salinity (high E.C. values), sensitivity to specific ions such as calcium, magnesium, sodium, chloride and perchlorate.

Salinity

Salinity damages may be caused by high E.C. ground water, irrigation water, soil or growth medium, and by excessive application of fertilizers.

High salt or E.C.: Leaves appear dull and leathery. A narrow yellow border develops around the leaf margin.

Plants grown under saline conditions are stunted and produce dark green, dull, leathery leaves that are prone to wilting. A narrow band of yellow necrotic tissue is often present on leaf edges. This can affect leaf expansion, causing a slight downward cupping of the leaf.

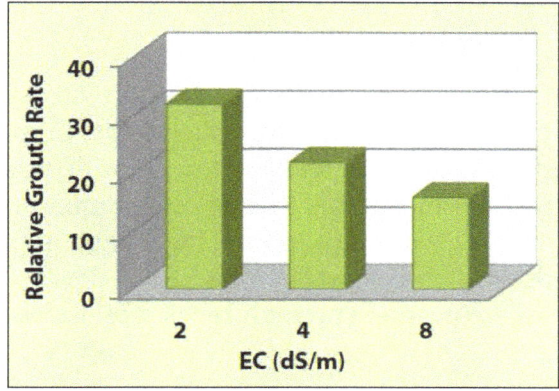

The effect of E.C. on relative growth rate of cucumber (cv. *Dina*) seedlings.

After a serious water stress, the oldest leaves may develop a uniform pale green chlorosis and small necrotic areas within the leaf. If water supply is maintained, leaves may only develop a band of pale green tissue around their edges. Plants are likely to wilt in warm weather.

Numerous studies have showed a linear decrease in the yields of cucumbers as the salt concentration of the irrigation water increased. The yield reduction was about 17% for a 1 mmho/cm increase in the E.C.

Figure, clearly shows also that cucumber plants are sensitive to saline conditions. Salinity hinders the vegetative growth of the plants. Moreover, salinity severely reduces crop productivity as shown in Figure.

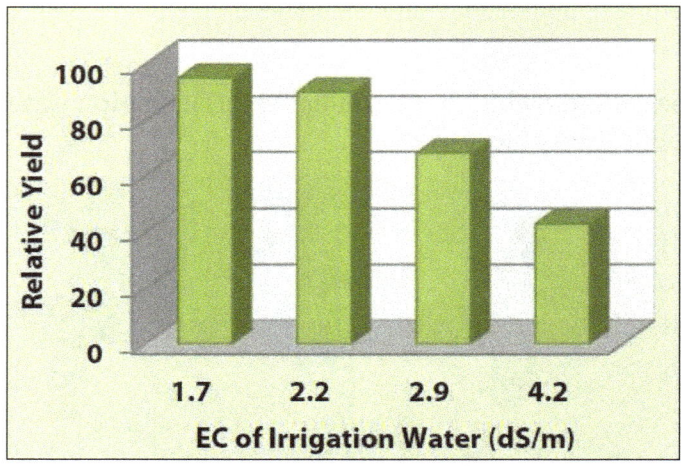

The effect of salinity on cucumber yield.

Cultural treatments to counteract salinity effects are:

- Leach the soil or growing medium with fresh water until the excess fertilizer is removed.

- Adjust the fertilizer program to ensure that rates do not exceed crop needs.

- Use cultivars that are more salinity resistant.

- Apply abundant amounts of potassium fertilizer, other than potassium chloride, as potassium markedly enhances plants ability to cope with salinity stresses.

Chloride (Cl⁻) Toxicity

Normal cucumber growth requires only small quantities of chloride similar to iron, but if the supply is plentiful more is taken up. Chloride toxicity can develop into a serious problem. Particular attention is required in recirculated hydroponic growth systems. Crop analysis showed that when sodium and chloride were added to the irrigation water, chloride was taken up in much greater quantities by the cucumber roots than sodium. The chloride anion markedly reduces plant vigor and tends to accumulate in the leaves margins, producing a band of pale green tissue around the leaf margin with some leaf-edge scorching and necrosis (tissue death), that stems from concentration rise up to 3% - Such leaves are prone to premature leaf abscission, and reduced photosynthesis activity.

Due to the specific chloride sensitivity of cucumbers, it is highly recommended to avoid fertilizers containing chloride, such as potassium chloride (muriate of potash (KCl), or calcium chloride ($CaCl_2$) in field grown cucumbers. Moreover, chloride-free fertilizers are an absolute prerequisite in hydroponic systems for best cucumber yields.

Chloride toxicity symptoms on cucumber leaf.

Perchlorate Toxicity

Perchlorate anion (ClO$_4$)$^-$ is present in mineral deposits of natural nitrates. Therefore, it is sometimes present in fertilizers produced from those deposits, like potassium nitrate produced in Chile. Perchlorate is a strong acid which decreases the activity of the enzyme RuDP (Ribulose Diphosphate Carboxylase).

When used for the cultivation of greenhouse vegetables it was proved that perchlorate at minute concentrations of 0.3% of the nutrient solution resulted in the *"Bolblad"* syndrome manifested by the following symptoms:

- Leaf curl and malformation developing into partial necrosis.

- Partial opening of the female flowers.

- Marked reduction in yields due to reduced fruit-set.

- Malformed cucumbers.

Glyphosate Toxicity

Glyphosate toxicity symptoms on cucumber leaves.

Glyphosate is a very popular wide-range herbicide. When used near cucumber plants, minute amounts absorbed by the cucumber plant are enough to produce damages that include upward-curling, pale green to yellow younger leaves and severe stunting.

Yields

Yields of cucumbers greatly vary according to varieties and growth conditions. The world average is 15 t/ha, while yields of 350 t/ha or more are obtained in modern greenhouses in Europe.

Beit Alpha varieties can yield: 250-450 t/ha and Dutch varieties yield 400 t/ha. Where yields are 30-60 t/ha under poor cultivation conditions, they can reach 100-300 t/ha under optimal temperature, humidity, light intensity and pollination, achieved under glass).

Harvesting

Cucumber crop matures within 40 - 50 days and harvesting starts 45 - 55 days after planting.

Harvest during summer to early fall depending on planting time, and variety. Unless an once-over mechanical harvester is being used, mainly for the pickling industry, fresh consumption cucumbers should be harvested at 2-4 day intervals, when the fruits have reached desired size, to avoid losses from oversized and over-mature fruit.

Over mature cucumbers left on the vine inhibit new fruit set. Pickling types are harvested when fruits are 5-7.5 cm (2-4 inches) long, and slicing types (for fresh market) - when fruits are dark green, firm, 15-20 cm (6-8 inches) long with a diameter of 4-5 cm (1½ - 2 inches).

Fresh-market cucumbers are ready for harvesting when they are about 6 to 10 inches long and 1.5 to 2.5 inches in diameter. The cucumber should be dark to medium green, without any signs of yellowing. On average, 58 to 65 days are required from seeding to maturity, depending on the cultivar and the growing conditions. Fresh-market cucumbers are harvested by hand. Because the individual fruits do not develop and mature consistently, the timing of maturity is not uniform within a field. As a result, fresh-market cucumbers typically are picked between 6 to 8 times over a 3-week period. In some situations, fresh-market cucumbers can be picked up to 12 to 15 times in a season. The number of pickings depends on when the seeds are planted and the supply and demand situation in the market. Price is an important factor in picking. Once prices for cucumbers fall below a certain level, it becomes uneconomical for growers to continue harvesting.

Pickling cucumbers are ready for harvesting when the ratio of length-to diameter ranges from 2.9 to 3.1. The cucumbers should be medium green, slightly lighter than fresh cucumbers, without any signs of yellowing. On average, it takes 55 to 65 days from seeding to maturity, depending on the cultivar and growing conditions.

Pickling cucumbers are either hand or machine harvested. Hand harvesting is common in areas where migrant labor is readily available. When harvested by hand, the field is typically picked 5 to 6 times. Pickling cucumbers must be harvested at 3- or 4-day intervals to prevent oversizing. If a machine is used for harvesting, however, the field is only picked once.

Eggplant

Brinjal or Eggplant is an important crop of subtropics and tropics. The name brinjal is popular in Indian subcontinents and is derived from Arabic and Sanskrit whereas the name eggplant has been derived from the shape of the fruit of some varieties, which are white and resemble in shape to chicken eggs. It is also called aubergine (French word) in Europe. The brinjal is of much importance in the warm areas of the Far East, being grown extensively in India, Bangladesh, Pakistan, China, and the Philippines. It is also popular in Egypt, France, Italy, and the United States. In India, it is one of the most common, popular and principal vegetable crops grown throughout the country except higher altitudes. It is a versatile crop adapted to dierent agro-climatic regions and can be grown throughout the year. It is a perennial but grown commercially as an annual crop. A number of cultivars are grown in India, consumer preference being dependent upon fruit color, size, and shape.

Brinjal Varieties

Below are the main brinjal hybrid varieties:

- Pusa Purple Long: It is an early maturing and long fruited type variety. Fruits are glossy, light purple in color, 25-30 cm long, smooth and tender. The crop is ready for picking in 100 to 110 days. Suitable for spring and autumn plantings, the average yield is 27.5 t/ha. It is moderately tolerant to shoot borer and little leaf disease.

- Pusa Purple Cluster: It is an early maturing and long fruited type variety. Fruits are small, dark purple in color and borne in clusters. The crop is ready for picking in 75 days after transplanting. Variety is resistant to little leaf disease under natural conditions.

- Pusa Kranti: This variety has a dwarf and spreading growth habit. Fruits are oblong and stocky than slender with attractive dark purple color. The crop matures in 130-150 days. Average yield is 14-16 t/ha.

- Pusa Barsati: This variety has a dwarf and erect growth habit devoid of thorns. Fruits are medium, long and purple with an average yield of 35.5 t/ha.

- Manjri Gota: This variety has a dwarf and spreading growth habit. The fruits are medium-large, round and purple colored with white stripes. Upon maturity, the fruits attain a golden yellow color. Average yield is 15- 20 t/ha.

- Vaishali: The variety has a dwarf and spreading type of growth habit. Fruits are oval in shape purple in color with white stripes. The stalks of the fruits bear spines. The crop is ready for rst picking within 60 days after transplanting. Average yield is 30 t/ha.

- Arka Navneet: This brinjal variety is high yielding hybrid. Fruits are large oval to oblong with deep purple shining skin with each fruit 450 g in weight. Purple owners with the solitary bearing habit. Free from bitter principles with very good cooking qualities. The crop is ready for picking in 150-160 days. Average yield is 65-70 t/ha.

- Arka Kusmukar: This brinjal type has spreading plant habit with green stem & green leaves. Flowers white green small fruits are borne in the cluster. Soft texture with good cooking quality. The crop is ready for picking in 140-150 days. Average yield is 40 t/ha.

- Arka Nidhi (BWR – 12): This is high yielding brinjal variety with resistance to bacterial wilt. Fruits are borne in the cluster. Calyx purplish green. Fruits free from bitter principles with slow seed maturity and good cooking quality. The crop is ready for picking in 150 days. Average yield is 48 t/ha.

- Arka Keshav (BWR – 21): This is a high yielding bacterial wilt resistant variety. Fruits tender, free from bitter principles with seed maturity. The crop is ready for picking in 150 days. Average yield is 45 t/ha.

- Arka Neelkanth (BWR – 54): This is a high yielding brinjal variety with bacterial wilt resistance. Fruits tender, free from bitter principles with seed maturity. The crop is ready for picking in 150 days. Average yield is 43 t/ha.

- Pusa Ankur: These types of brinjal are oval-round, small-sized (60-80g), dark purple, attractive fruits. Fruits are small, oval-round, bark purple, glossy and very attractive, weighing each 60-80g. It is an early bearing and becomes ready for rst picking 45 days after transplanting. Its fruits do not lose color and tenderness even on delayed pickings.

1. The Climatic Requirement for Brinjal Production: The brinjal is a warm season crop, therefore susceptible to severe frost. Low temperature during the cool season causes deformation of vegetables. A long and warm growing season is desirable for successful brinjal farming. Cool nights and short summers are not suitable for satisfactory production. A daily average temperature of 13 to 21 °C is most favourable for optimum growth and yield. The brinjal seed germinates well @ 25 °C.

2. Soil Requirement of Brinjal Plantation: The brinjal plants can be grown in all types of soil varying from light sandy to heavy clay. Well-drained soil is rich in organic matter with a pH of 6.5-7.5. Light soils are good for an early yield, while clay-loam and silt-loam are well suited for higher yield.

3. Soil Preparation for Brinjal Plantation: Since the crop remains in the eld for a number of months. The soil should be thoroughly prepared by ploughing 4 to 5 times before transplanting the seedlings. Bulky organic manures like well rotten crowding or compost should be incorporated evenly on the soil. Thoroughly prepare the eld with the addition of FYM @ 25 tonnes/ha and form ridges and furrows at a spacing of 60 cm. Apply 2 kg/ha of Azospirillum and 2 kg/ha of Phosphobacteria by mixing with 50 kg of FYM. Irrigate the furrows and transplant 30-35 days old seedlings at 60 cm apart on the ridges.

4. Season of Sowing for Brinjal seeds: December – January, and May – June.

5. Nursery bed preparation for Brinjal seedlings: Apply FYM 10 kg, neem cake 1 kg, VAM 50 g, enriched superphosphate 100 g and furadon 10 g per square meter before sowing. Area required for raising seedling for planting 1.0 ha is 100 sq.m.

6. A seed rate of Brinjal: Varieties : 400 grams /ha, Hybrids : 200 grams/ha.

7. Brinjal seed treatment procedure: Treat the brinjal seeds with Trichoderma viride @ 4 g / kg or Pseudomonas uorescens @ 10 g / kg of seed. Treat the seeds with Azospirillum @ 40 g / 400 g of seeds using rice gruel as adhesive. Irrigate with a rose can. In raised nursery beds, sow the seeds in lines at 10 cm apart and cover with sand. Transplant the seedlings 30 – 35 days after sowing at 60 cm apart in the ridges.

The eggplant seeding.

Transplanting in Brinjal Farming: The seedlings are ready in 4-5 weeks for transplanting, when they attained a height of 12-15 cm with 3- to 4 leaves. Harden the seedlings by withholding irrigation. Uproot the seedlings carefully without injury to the roots. Transplanting should be done during evening hours followed by irrigation. Firmly press the soil around the seedlings. Spacing depends upon the fertility status of soil, type of verities and suitability of the season. In general 60×60 cm spacing is kept for non-spreading type verities and 75- 90×60-75 cm for spreading type verities.

Water requirement of Brinjal plants: Water the eld as per the need of the crop. Timely irrigation is quite essential for good growth, powering, fruit setting and development of fruits. Higher yield may be obtained at optimum moisture level and soil fertility conditions. In plains, irrigation should be applied every third to the fourth day during hot weather and every 7 to12 days during winter. Irrigation is given before top dressing of there is no rain. The brinjal eld should be regularly irrigated to keep the soil moist during frosty days.

Layout and planting for drip irrigation and fertigation of Brinjal:

- Apply FYM @ 25 t / ha as basal dose before the last plowing.

- Apply 2 kg/ha of Azospirillum and 2 kg/ha Phosphobacteria by mixing with 50 kg of FYM.

- Apply 75 % total recommended dose of superphosphate i.e. 703 kg/ha as basal.

- Install the drip irrigation with main and sub-main pipes and place lateral tubes at an interval of 1.5 m.

- Place the drippers in lateral tubes at an interval of 60 cm and 50 cm spacing with 4 LPH and 3.5 LPH capacities respectively.

- Form raised beds of 120 cm width at an interval of 30 cm and places the laterals at the center of each bed.

- Before planting, wet the beds using a drip system for 8-12 hrs.

- Planting to be done at a spacing of 90 x 60 x 75 cm in the paired row system, using ropes marked at 75 cm spacing.

- Spray Pendimethalin 1.0 kg a.i./ha or Fluchloralin 1.0 kg a.i/ha as pre-emergence herbicide at 3rd day after planting.

- Gap lling to be done at 7th day after transplanting.

Intercultural Operations and Weed Control of Brinjal Crop

The weeds should be controlled as soon as they have seen, either by the traditional method of hand weeding and hoeing or by application of herbici8des. Regular or frequent shallow cultivation should be done at regular intervals so as to keep the field free from weeds and to facilitate soil aeration and proper root development. The most serious weed in brinjal is the *Orabanchae sp*. It is root parasite and it should be controlled carefully. Gap filling should be done wherever needed during evening hours followed by irrigation. Pre-plant soil incorporation of Fluchloralin (1- 1.5 kg/ha) or Oxadiazon (0.5 kg/ha) and pre-planting surface spraying of Alachlor (1-1.5 kg/ha) control the weeds of brinjal successfully. Manuring application for Brinjal crop: Apply 2 kg each of Azospirillum and Phosphobacteria in the main field at planting.

- Varieties: Basal dose: FYM 25 t/ha, NPK 50:50:30 kg/ ha. Topdressing: 50 kg N/ha on the 30th day of planting or during earthing up.

- Hybrids: Basal dose: FYM 25 t/ha, NPK 100:150:100 kg/ha. Topdressing: 100 kg N/ha on the 30th day of planting or during earthing up.

Fertigation Schedule for Hybrids of Brinjals

Age	Crop stage	Duration in days	Fertilizer grade	Total Fertilizer (kg/ha)	Nutrient applied			% of requirement		
					N	P	K	N	P	K
1	Transplanting to plant establishment stage	10	19:19:19 +MN 13:0:45 Urea	39.47 5.50 25.65	7.50 0.70 11.80	7.50 – –	7.50 2.50 –	10.00	5.00	10.0
				Subtotal	20.00	7.50	10.00			
2	Vegetative stage	30	12:61:0 13:0:45 Urea	24.50 88.89 142.4	2.94 11.56 65.50	15.00 – –	40.00 – –	40.00	10.00	40.0
				Subtotal	80.00	15.00	40.00			
3	Flower initiation to first picking	30	19:19:19 +MN 13.0:45 Urea	39.47 50.00 100.00	7.50 6.50 46.00	7.50 – –	7.50 22.50 –	30.00	5.00	30.0
				Subtotal	60.00	7.50	30.00			
4	Harvesting	80	12:61:0 13:0:45 Urea	12.30 44.40 71.13	1.48 5.80 32.72	7.50 – –	– 20.00 –	20.00	5.00	20.0
				Subtotal	40.00	7.50	20.00			
					200.00	37.50	100.00	100	25	100

Recommended dose: 200:150: 100 kg/ha. 75% of RD of Phosphorus applied as superphosphate = 703 kg/ha.

1. 19:19:19 = 79 kg/ha.

2. 13:0:45 = 189 kg ha.

3. 12:61:0 = 37 kg/ha.

4. Urea = 340 kg/ha.

Pests and Diseases of Brinjal Plant

Shoot and Fruit Borer

- Remove the affected terminal shoot showing boreholes.

- Remove the affected fruits and destroy.

- Avoid using synthetic pyrethroids.

- Spray neem seed kernel extract 5% or any one of the following chemicals starting from one month after planting at 15 days interval.

Insecticide	Dose
Azadirachtin 1.0% EC (10000 ppm)	3.0 ml/lit.

Azadirachtin 0.03 % WSP (300 ppm)	5.0 g/lit.
Chlorpyrifos 20 % EC	1.0 ml/lit.
Dimethoate 30 % EC	7.0 ml/10 lit.
Emamectin benzoate 5 % SG	4 g/10 lit.
Flubendiamide 20 WDG	7.5 g/10 lit.
Phosalone 35 % EC	1.5 ml/lit.
Quinalphos 20 % AF	1.7ml/ lit.
Quinalphos 25 % EC	1.5 ml/lit.
Thiodicarb 75 % WP	2.0 g/lit.
Thiometon 25 % EC	1.0 ml/lit.
Trichlorofon 50 % EC	1.0 ml/lit.
Triazophos 40 % EC	2.5 ml/lit.

- Damping off: Treat the seeds with *Trichoderma viride* 4 g/kg or *Pseudomonas fluorescens* 10 g /kg of seed 24 hours before sowing. Apply *Pseudomonas fluorescens* as soil application @ 2.5 kg/ha mixed with 50 kg of FYM. Water stagnation should be avoided. Drench with Copper oxychloride at 2.5 g/lit at 4 lit/sq.m

- Leaf Spot: Leaf spot can be controlled by spraying Mancozeb 2 g/lit in brinjal farming.

- Little Leaf: Remove the affected plants in the early stages and spray Methyl demeton 30 EC @ 1.0 ml/lit. to control the vector.

Harvesting and the Yield of Brinjal

The brinjal fruits are harvested when they attain full size and color but before the start of ripening. Tenderness bright color and glossy appearance of the fruit is the optimum stage of harvesting of fruits. When the vegetables look dull, it is an indication of maturity and loss of quality in brinjal farming.

Harvested Brinjal.

The yield varies from season to season, variety to variety and location to location. However, in general, 250 to 500 quintals/ha of healthy vegetables of brinjal can be obtained from your brinjal farming.

Pumpkin

Pumpkin (*Cucurbita moschata*) is a popular vegetable of Kerala. It is a rich source of potassium and Vitamin A. The bright orange color of pumpkin is an indication of an important antioxidant, beta carotene. Beta-carotene is the precursor of vitamin A in the body, which performs many important functions in overall health. The name pumpkin is originated from "pepon" the Greek word for "large melon." Central America is the centre of origin of pumpkin.

Climate and Soil

Pumpkin requires a minimum temperature of 18 °C during early growth, but optimal temperatures are in the range of 24–27 °C. It can tolerate low temperatures and are adapted to a wide variety of rainfall conditions. Pumpkin tolerates a wide range of soil but prefers a well drained sandy loam soil that is rich in organic matter. The optimum soil pH is 6.0–6.7, but plants tolerate alkaline soils up to pH 8.0.

Varieties

- Arka Suryamukhi: Small sized, flat-round fruits with orange colour. Resistant to fruit fly attack. Suitable for growing in Kerala condition during September-January.

- Arka Chandan: Fruits are round with pressed blossom end. Rind colour green with white patches when immature which turns to light brown. upon maturity. Fruits are with thick orange flesh and rich in carotene, and has solid cavity. Fruit weight is 2-3 kg. Average yield is 33 t/ha. Crop duration is 115-120 days.

- Ambili: High yielding variety released from the Kerala Agricultural University. Average fruit weight is 4-6 kg. Medium sized flat round fruits.

- Saras: A medium sized pumpkin variety with attractive flesh colour and more flesh content, released from the Kerala Agricultural University. Average yield is 39 t/ha. Highly suited for growing in Thrissur, Palakkad and Ernakulam districts.

- Suvarna: High yielding variety released from the Kerala Agricultural University. Small sized flat- round fruits with thick orange flesh. Average fruit weight is 3-5 kg.

Propagation and Planting

- Seed rate: Approximately 1.0-1.5 kg of seeds are required for cultivating one hectare of land.

- Planting: January-March and September-December are the ideal seasons for growing pumpkin. For the rain fed crop, sowing can be started after the receipt of first few showers during May-June. Prepare the soil to a fine tilth by ploughing and pits of 60 cm diameter and 30-45 cm depth are taken at a spacing of 4.5 x 2 m. Well rotten FYM and fertilizers are mixed with topsoil in the pit.

Sow four or five seeds in a pit at 1-2 cm depth. Deeper sowing delays germination. As seedlings require ample water for quicker germination, a pre-sowing irrigation 3-4 days before sowing is beneficial. Irrigate with a rose can daily. The seeds germinate in about 4-5 days. Unhealthy plants are removed after two weeks and only 3 plants are retained per pit.

Intercultural Operations

- Trailing: Pumpkin grows very fast and vines elongate rapidly within two weeks after planting. Thereafter, the plant sends out lateral stems. Usually, pumpkin is grown trailing on the ground. For trailing them, spread dried twigs on the ground.

- Manuring: Balanced fertilization is essential for high yielding and good keeping quality of the fruits. Apply FYM @ 20-25 t/ha as basal dose along with half dose of N (35 kg) and full dose of P_2O_5 (25 kg) and K_2O (25 kg/ha). The remaining dose of N (35 kg) can be applied in two equal split doses at the time of vining and at the time of full blooming. A fertilizer dose of 70:25:25 kg $N:P_2O_5:K_2O$/ha in several splits is recommended in Onattukara region. The fertilizer dose per pit would be 28:10:10 g $N:P_2O_5:K_2O$.

- Irrigation: During the initial stages of growth, irrigate at 3-4 days interval, and alternate days during flowering/fruiting. Furrow irrigation is the ideal method of irrigating. But in water-limited environment, trickle or drip irrigation can be resorted to. During rainy season, drainage is essential for plant survival and growth.

Weed Control

Conduct weeding and raking of the soil at the time of fertilizer application. Earthing up is done during rainy season. Hand or hoe weeding can be performed as needed. Mulching is commonly used for pumpkin crops grown on raised beds. Use organic or plastic mulch depending on availability. Mulch can be laid down before or after transplanting and after sowing.

Pests

- Fruit Flies: Bactocera Cucurbitae-Fruit fly is the most destructive insect pest of pumpkin. Fruit fly maggots feed on the internal tissues of the fruit causing premature fruit drop and also yellowing and rotting of the affected fruits. This fly is difficult to control because its maggots feed inside the fruits, protected from direct contact with insecticides.

  ○ Control: Bury any infested fruits to prevent the build up of fruit fly population. In homestead gardens, covering the fruits in polythene/paper covers help to prevent flies from laying eggs inside the fruits. Breaking of soil to expose pupae, and burning the soil in pit by dried leaves are also effective. It can also be effectively controlled by the use of banana fruit traps.

- Epilachna Beetle: Epilachna Spp-The yellowish coloured grubs and adults of the beetle feed voraciously on leaves and tender plant parts, and the leaves are completely skeletonized leaving only a network of veins. When in large number, the pest causes serious defoliation and reduces yield.

  ○ Control: Remove and destroy egg masses, grubs and adults occurring on leaves. Spray carbaryl 0.2%.

- Pumpkin Beetle: Aulacophora Fevicolis, A. Cincta And A. Intermedia-Adult beetles eat the leaves, makes hole on foliage and causes damage on roots and leaves. Grubs cause damage by feeding on root. It also feeds on flowers and bores into developing fruits that touch the soil.

  ○ Control: Remove and destroy egg masses, grubs and adults occurring on leaves. Spray carbaryl 0.2%.

- Pumpkin Beetle: Aulacophora Fevicolis, A. Cincta And A. Intermedia-Adult beetles eat the leaves, makes hole on foliage and causes damage on roots and leaves. Grubs cause damage by feeding on root. It also feeds on flowers and bores into developing fruits that touch the soil.

  ○ Control: Incorporate carbaryl 10% DP in pits before sowing the seeds to destroy grubs and pupae.

- Aphids: Aphis Gossypi-Aphids in large number congregate on tender parts of plant and suck sap resulting in curling and crinkling of leaves. Ants carry aphids from one plant to another.

  ○ Control: Apply 1.5% fish oil soap. First dissolve soap in hot water and then make up the volume. Alternatively apply dimethoate 0.05%.

Diseases

- Downy Mildew: Pseudoperonospora Cubensis-Cottony white mycelial growth is seen on the leaf surface. Chlorotic specks can be seen on the upper surface of the leaves. It is severe during rainy season.

  ○ Control: Complete removal and destruction of the affected leaves. Spraying 10 % solution of neem or kiriyath preparation. If the disease incidence is severe spraying mancozeb 0.2% will be useful.

- Powdery Mildew: Erysiphe Cichoracearum-The disease appears as small, round, whitish spots on leaves and stems. The spots enlarge and coalesce rapidly and white powdery mass appears on the upper leaf surface. Heavily infected leaves become yellow, and later become dry and brown. Extensive premature defoliation of the older leaves resulting in yield reduction.

  ○ Control: Control the disease by spraying Dinocap 0.05%.

Mosaic (Cucumber Mosaic Virus)

- Mosaic disease is characterized by vein clearing and chlorosis of leaves. The yellow network of veins is very conspicuous and veins and veinlets are thickened. Growths of plants infected in the early stages remain stunted and yield of the plant get severely reduced. White fly (Bemisia tabaci) is the natural vector of this virus.

  ○ Control: Control the vectors by spraying dimethoate 0.05%. Uprooting and destruction of affected plants and collateral hosts should be done.

Harvesting

Pumpkins are ready to harvest when the stems connecting the pumpkin to the vine begin to shrivel. Harvest the fruits whenever they are a deep, solid color (orange for most varieties) and the rind is hard. Pumpkins that are not fully mature or that have been injured do not store well. Cut pumpkins from the vines carefully, using pruning shears or a sharp knife leaving 3-4 inches of stem attached. Snapping the stems from the vines results in fruits without stem attached, which reduce the storage life.

The crop yield can vary depending on variety and crop management. Average marketable yields are 30 t/ha. Avoid cutting and bruising the pumpkins when handling them. Wash and dry the harvested fruits thoroughly, and cure for several days in sunlight before storing them. The harvested fruits can be stored for several weeks in ambient conditions. Pumpkins will keep for 2-3 months in temperatures from 10 to 12 °C and 50-75 % relative humidity.

Guava

Guava Fruit (*Psidium guajava*) is one of the most common fruits in India. It is the quite hardy and prolific bearer. Guava is a commercially significant, highly remunerative crop even without much care. It is a rich source of vitamin C and pectin. It is also a good source of calcium and phosphorus.

Certain important strategies have been identified for enhancing horticulture development in India in order to be competitive in the world market. They involve the adoption of modern, innovative and hi-tech methods. One such strategy is the high-density plantation (HDP). This includes the adoption of appropriate plant density, canopy management, quality planting material, support, and management system with appropriate inputs. HDP generally refers to planting at a closer spacing than the normally recommended spacing. It has been attempted in different crops such as guava, apple, banana, mango, pineapple, peach, etc. Many guava farmers have been adopting this technology successfully in different parts of the country. HDP technology results in maximization of unit area yield and availability of the fruits in the market early which fetch a better price.

Suitable Location

Guava orchard

Guava Fruit is successfully grown all over India. The total area and production of guava in the country are 1.90 lakh hectare and 1.68 million tonnes. Major guava producing states are Bihar, Uttar Pradesh, Maharashtra, Karnataka, Orissa, West Bengal, Andhra Pradesh, and Tamil Nadu. However, Uttar Pradesh is by far the most important guava producing state of the country and Allahabad has the reputation of growing the best guava in the country as well as in the world.

Required Soil

Guava Fruit is very hardy. It can thrive on all types of soil from alluvial to lateral. However, it is sensitive to waterlogging. It can be grown on heavier but well-drained soil. Deep friable and well-drained soils are the best. The topsoil should be rich for a better stand. Soil pH range of 4.5 to 8.2 is congenial for guava but saline or alkaline soils are unsuitable.

Required Climatic Conditions

Guava Fruit is successfully grown under both tropical and subtropical climates. It can grow from sea level to an altitude of about 1500 m (5000'). Annual rainfall of below 1000 mm (40') between June and September is the best for the growth of guava plants. Young plants are susceptible to drought and cold conditions. Yield and quality improvement in areas with a distinct winter season.

Guava Fruit Varieties

The most popular guava Fruit cultivars are Lucknow 49, Allahabad Safeda and Harijha. Other varieties preferred by the farmers are Apple, Baruipur Local, Benarasi, etc. From the viewpoint of yield and quality, Lucknow-49 may be considered to be the most popular commercial cultivar. Different research institutes have been making efforts to develop some new varieties and hybrids. IIHR, Bangalore, has developed two soft-seeded superior varieties viz., Arka Mridula and Arka Amulya.

Guava Tree Propagation

Guava fruit farming.

Guava is propagated from seeds and also by vegetative methods. Seedling trees produce fruits of variable size and quality although such trees are generally long-lived. Vegetative methods like cutting, air layering, grafting, and budding are used for propagation of guava. Air-layering has been observed to be the most successful commercial method practiced for guava. The cheapest method of rapid multiplication is stooling, i.e. mound layering in nursery beds.

Cultivation Technology in Guava Fruit Farming

Planting Method of Guava

The field should be deeply plowed, cross plowed, harrowed and levelled before digging pits. The pits of about 0.6 m x 0.6m x0.6 m dimension should be dug before the monsoon. After 15-20 days, each pit should be filled with soil mixed with 20 kg of organic manure and 500 g of superphosphate. In very poor soils, the pit size may be bigger, about 1m x 1m x 1m, and more of organic manures may be necessary. The onset of monsoon is the time to start planting.

Planting Density in Guava Farming

Standard spacing for guava is, 6m x 6m, accommodating 112 plants /acre. However, it is commonly planted at a distance of 3.6 m to 5.4m (12′ to 18′). Traditional planting spaces in some parts of the country range even up to 5.4 to 7.0m (18′ to 23′). By increasing plant density, productivity can be increased. Although there would be a reduction in the size of fruits, the number of fruits per plant remains more or less similar. In the model scheme, a distance of 4.5m x 4.5m (15'x15′) with a population of 195 per acre is considered, which was observed to be common in areas covered during a field study.

Irrigation for Guava Plants

Normally irrigation is not required in guava plantation. However, in the early stage, young guava plants require 8 to 10 irrigations a year. Life-saving hand watering is necessary for the summer season in dry areas and on light soils. Full grown bearing trees require watering during May-July at weekly intervals. Irrigations during winter reduce fruit drop and improve fruit size of winter crop. In order to conserve soil moisture from pre-monsoon showers, V-shaped or half- moon shaped bunds or saucer-shaped basins may be made. Drip irrigation has been proved to be very beneficial for guava. Besides saving 60 % of water, it results in a substantial increase in size and number of fruits

Manuring and Fertilization in Guava Orchard

Guava is very responsive to the application of inorganic fertilizers along with organic manures. Soil type, nutrient status, and leaf analysis can give a better indication of the requirement of nutrients. A thumb rule recommendation is considered in this model. NPK may be applied @100, 40 and 40 g per plant year of age, with stabilization in the 6th year. They may be applied in two equal split doses in January and August.

Spraying the trees with 0.45 kg zinc sulfate and 0.34 kg slaked lime dissolved in 72.74 l (16 gallons) of water cures Zn deficiency. The number of sprays depends on the severity and extent of the deficiency. Pre-flowering sprays with 0.4% Boric Acid and 0.3% Zinc Sulphate increase the yield and fruit size. Spraying of copper sulfate at 0.2 to 0.4% also increases the growth and yield of guava.

Intercultural Operations of Guava Plants

The main practices of intercultural operations followed are weeding and spading. Manual weeding is preferable; spraying weedicides such as gramoxone is also effective. in order to manage the orchard soil, plowing two times a year, once in October and the other in January, is necessary. Mulching the basins at least twice a year also is important to conserve moisture and discourage weed growth.

Intercropping in Guava Fruit Farming

The interspace can be economically utilized by growing suitable intercrops in the early stages till the bearing. A crop combination of several plantation crops, vegetables and leguminous crops like papaya, pineapple, beans, cucumber, cabbage, cauliflower, peas, cowpea, etc., are considered safe intercrops.

Training and Pruning of Guava Plant

Training of guava trees improves yield and fruit quality. The main objective of training guava plants is to provide a strong framework and scaffold of branches suitable for bearing a heavy remunerative crop without damaging the branches. For this, shoots coming out close to the ground level should be cut off up to at least 30 cm from the soil. The center should be kept open, while four scaffold limbs may be allowed to grow. Light annual pruning is necessary for guava as it bears on current season's growth. Experimental evidence supports pruning off 75% of current season's growth in May for harvesting good winter crop.

Pest Control Management of Guava Plants

The fruit fly, mealy bug, scale insects, etc. are the major pests in guava. The following measures are adapted to control the damage done by these pests:

- Fruit fly: Spraying of chemicals like malathion 2 ml, phosphamidon 0.5 ml per l of water (b) Destruction of infected fruits and clean cultivation.

- Mealybug: Soil treatment with Aldrin, malathion, thimet, ete (b) Banding the base of the plant with polythene film to prevent the nymph from climbing up from the soil. (c) Spraying of methyl parathion, monocrotophos or dimethoate.

- Scale insect: Spraying of fish oil rosin soap with water or crude oil emulsion,dimetholate, and methyl demiton, etc.

Disease Control Management of Guava Plants

The most damaging diseases in guava are wilt and anthracnose. Canker, cercospora leaf spot, seedling blight. etc., are some other important diseases. Control measures of the major diseases are briefed below:

- Wilt disease: Wilt is the most serious fungal disease. Bearing trees, once affected, slowly die away. Drenching the soil at trunk bases with brasicol and spraying the plant with bavistin at an early stage of infection minimize the damage. Injecting 8-Quinolonol sulfate is also effective.

- Anthracnose: Spraying of Cu-oxychloride, cuprous oxide, difolatan, dithane Z- 78, etc., control this disease.

Flowering and Fruit Set in Guava

Two important seasons of blooming are observed, one in April-May (Monsoon Crop) and the other in September – October (Winter Crop). Growth regulators like NAA, NAD, and 2,4-D are very effective in thinning of flowers and manipulating the cropping season. Fruit drop in guava is as severe as 45-65% due to different physiological and environmental factors. Spraying of GA is highly effective in reducing the drop.

Guava Fruit Harvesting

Grafted, budded or layered guava trees start bearing at the age of 2 to 3 years. Seedling trees require 4 to 5 years to bear. The guava fruit cannot be retained on the tree in the ripe stage. So, it should be picked immediately when it is mature. Guava is ready for harvest as soon as the deep green color turns light and a yellowish green patch appears. Individual hand picking at regular intervals will avoid all possible damage.

The Yield of Guava Fruit

The yield varies in different cultivars and with care and management of the orchard, age of plant and season of cropping. The yield per tree may be as high as 350 kg from grafted plants and 90 kg from the seedling tree. A three-year-old grafted Lucknow – 49 guava tree may yield 55-60 kg under suitable conditions. Yield starts with 4 to 5 kg in the second year.

Crop Regulation of Guava

Compared to monsoon crop, winter crop is much superior in quality and fetch a premium price. Therefore, farmers often reduce monsoon crop by deblossoming to get a higher price. This is done by spraying plant regulators like Maleic Hydrazide (100000 ppm) on spring flush of flowers. NAA 100 ppm, NAD 50.ppm, or 2,4-D 30 ppm are also reported to be effective in thinning flowers. Root exposure and root pruning are done to bring flowers at the desired time. Sometimes bending of twigs is done to force new sprouts which come up with flowers. Hand thinning of flowers is also very effective. Defoliation is also recommended sometimes for forcing new growth with flowers.

Post-harvest Management of Guava Trees

Fruit yield

Guava is highly perishable in nature. Shelf life under ambient conditions is 2 to 3 days on an average. Therefore, it should be marketed immediately after harvest. However, it may be stored for a few days to adjust the market demand. After careful harvest, the fruits should be brought to the packhouse. For packing, corrugated fiberboard with adequate perforation may be used. However, fruits are reported to keep 3 to 5 weeks in cold storage at a temperature of 8 to 10 degree Celsius with 85 -90 % RH.

Banana

Banana is one of the major and economically important fruit crop of Asian countries. Banana occupies vast area among the total area under crop cultivation in Asia region. Bananas are the fourth largest fruit crop in the entire world and most of Banana is cultivated by planting suckers. As technology development in agriculture is very fast, it results in developing tissue culture technique. Growing bananas does not require much eort but to achieve high yields or production requires dedication, farm management skills, and proper planting methods. Banana plant belongs to the family of "Musaceae" and genus of "Musa". Bananas are indigenous to the tropical portions of India, Southeast Asia and northern Australia. Basically banana plants are not trees but giant herbs, which will reach their full height of between 10 feet and 20 feet after only a year. Every banana blossom develops into a fruit and ripe enough for consumption after about 4 to 5 months. After producing banana fruit, the plant stems die o and they will be replaced by new growth. The number of bananas produced by each plant varies based on fruit variety and other factors. Apart from being consumed as a fresh fruit, Banana leaves are used worldwide as cooking materials, plates, umbrellas, seat pads for benches, shing lines, clothing fabric.

- Varieties of Banana: There are many varieties of banana grown across Asia. However, some of the popular varieties of banana are Red banana, Nyali, Safed Velchi, Basarai, Ardhapuri, Rasthali, Karpurvalli, Dwarf Cavendish, Robusta, Monthan, Poovan, Nendran, Grandnaine, Karthali, Dwarf Cavendish, Robusta, Monthan, Poovan, Nendran, Red banana, Nyali, Safed Velchi, Basarai, Ardhapuri, Rasthali, Karpurvalli, Karthali and Grandnaine, Emas, Rastali, Raja Awak, Abu, Nangka and Tanduk. Out of all these, Grandnaine is gaining popularity and may soon be the most preferred cultivar due to its tolerance to biotic stresse and good quality of bunches.

- Climate Requirement for Banana Farming: Banana is basically a tropical crop, grows well in temperature range of 14 °C to 38 °C with RH regime of 75% to 85%. Bananas need warm climate, adequate moisture and protection from wind. Chilling injury occurs at temperatures below 12 °C. The normal growth of the banana begins at 18 °C, reaches optimum at 28 °C, then declines and comes to a halt at 38 °C. Although Bananas grow best in bright sunlight, higher temperature causes sun scorching. High velocity winds which exceed 80 km/hr damage the banana crop.

- Soil Requirement for Banana Farming: Banana can be cultivated on wide range of soils. However, rich, moisture and well-drained soils with 45% clay, 70% silt, 80% loam are best for its growth. Banana plants prefer a more acidic soils with pH between 6.5 to 7.5. Soils with low pH value make banana plants more susceptible to Panama disease. Sandy, salty, nutritionally decient and ill-drained soil, low laying areas, black cotton soils with poor drainage should be avoided for banana farming. Supplement the soils which are decient in nutrients with organic matter before planting the Banana trees. Banana plants require thick mulching for retaining the water and this process should be repeated as often as possible. Banana plants are very sensitive to waterlogging, because its roots will rot. This however can be resolved by planting the banana in raised beds. If you are planning for commercial banana farming, it is advised to go for soil test.

- Propagation in Banana Farming: Generally, Propagation in Banana Farming is done by suckers or tissue culture plants.

- Land Preparation in Banana Farming: Growing green manuring crops like cowpea or daincha and burying it in the soil before the planting the banana is benecial. The main eld should be levelled and make weed free by give 3 to 4 ploughings and using harrow or rotavator or any suitable agriculture equipment to bring the soil to ne tilth stage. During the nal plough, apply well rotten farmyard manure (FMY) as a basal dose in the soil and make sure it will be mixed well into the soil. A required pit size of $45 \times 45 \times 45$ cm should be dug. These pits should be re-lled with topsoil along with welldecomposed farm yard manure of (FMY), 250 grams of Neem cake, and 20 grams of conbofuron. For better aeration and to prevent any soil borne diseases, these prepared pits should be exposed to sunlight for some time. This also kills any harmful insects. In case of saline or alkali soils having pH above 8.0, the pit mixture should be modied to incorporate organic matter.

Adding more organic manure will reduce the salinity. Alternatively, planting can be done in furrows.

Growing Bananas from Seeds.

- Planting Material in Banana Farming: Suckers weighing approximately 500 to1000 grams are commonly used in propagating. The best way is to start with tissue culture plantlets. Tissue culture plantlets are recommended for planting because suckers, in general, are infected with some soil-borne pathogen and nematodes. Similarly due to the variation in age and size of sucker the crop will not be uniform, harvesting of the crop will be prolonged and crop-management becomes very dicult.

Apple

Apple (Malus pumila) is an important temperate fruit. Apples are mostly consumed fresh but a small part of the production is processed in to juices, jellies, canned slices and other items.

Climate

The apple is a temperate fruit crop. However, the apple growing areas do not fall in temperate zone but the prevailing temperate climate of the region is due to the Himalayan ranges and high altitudes. The average summer temperature should be around 21-24 °C during active growth period. Apple succeeds best in regions where the trees experience uninterrupted rest in winter and abundant sunshine for good colour development. It can be grown at an altitude of 1500- 2700 m above the sea level. Well-distributed rainfall of 1000-1250 mm throughout the growing season is most favourable for optimum growth and fruitfulness of apple trees.

Soil

Apples grow best on a well-drained, loam soils having a depth of 45 cm and a pH range of pH 5.5-6.5. The soil should be free from hard substrata and water-logged conditions. Soils with heavy clay or compact subsoil are to be avoided.

Propagation

- Grafting: Apples are propagated by several methods viz.; whip, tongue, cleft and roots grafting. Tongue and cleft grafting at 10-15 cm above the collar during February-March gives the best results. Usually grafting is done at the end of winter.

- Budding: Apples are mostly propagated by shield budding, which gives a high percentage of success. In shield budding a single bud along with a shield piece of stem is cut along with the scion and inserted beneath the rind of the rootstock through a 'T' shaped incision during active growth period. Budding is done when the buds are fully formed during summer. The optimum time of budding is September in Kashmir Valley, Kumaon hills of Uttaranchal, high hills of Himachal Pradesh and June in mid hills of Himachal Pradesh.

- Rootstocks: Most of the apple plants are grafted or budded on seedling of wild crab apple. The seedling rootstocks obtained from the seeds of diploid cultivars like Golden Delicious, Yellow Newton, Wealthy, Macintosh and Granny Smith also can be used. High density planting is done using dwarfing rootstocks (M9, M4, M7 and M106).

Planting

The planting distance varies according to variety and the fertility level of the soil. The main consideration in planting trees is planting of sufficient pollinators to ensure effective pollination. Usually one pollinator tree is needed for two to three large trees planted at 10 m distance or one row pollinator for two rows of main cultivar. For high density planting the pollinator tree is planted after every sixth tree in a row.

The most widely used planting system is the square system. In this system, the pollinators are planted after every sixth or ninth tree. The other popular system of planting is the rectangular system. In hilly areas the apple orchards are established by planting the trees on the contours so as to prevent soil erosion and reduce run off.

The average number of plants in an area of one ha. can range between 200 to 1250. Four different categories of planting density are followed viz. low (less than 250 plants/ha.), moderate (250-500 plants/ha.), high (500-1250 plants/ha.) and ultra-high density (more than 1250 plants /ha.). The combination of rootstock and scion variety determines the plant spacing and planting density/unit area.

Time and Method of Planting

Planting is usually done in the month of January and February. Pits measuring 60 cm are dug two weeks before planting. The pits are filled with good loamy soil and organic matter. Planting is done in the centre of the pit by scooping the soil and placing the soil ball keeping the roots intact. Loose soil is filled up in the remaining area and lightly pressed to remove air gaps. The seedlings are staked and watered immediately.

Irrigation

Apple trees are particularly sensitive to low soil moisture. Water stress during the growing season reduces number and size of fruits, and increases June drop. Success of apple largely depends on uniform distribution of rain during the year in case of dry spells during the critical periods supplementary irrigation should be provided. Water stress conditions results in poor fruit set, heavy fruit drop, low production and poor quality. The most critical periods of water requirement are April- August and peak water requirement is after fruit set. Normally the orchards are irrigated immediately after Manuring in the month of December-January. During the summer periods, the crop is irrigated at an interval of 7-10 days. After the fruit setting stage the crop is irrigated at weekly intervals. Application of water during the fortnight preceding harvest markedly improves the fruit colour. Thereafter till the onset of dormancy, irrigation is given at an interval of 3-4 weeks.

Pruning

Pruning is one of the most important practice which promotes plant vigour and productivity. Pruning is done with a view to divert the sap flow towards the fruiting branches and to force the plants to bear more fruits or to induce vigorous vegetative growth. During pruning, weak-growing and diseased branches are removed from the tree. Usually the trees are pruned every year in the month of December-January. The systems of pruning adopted in apple cultivation are as follows.

- Established Spur System: Objective of this pruning is to develop permanent fruit spurs for production of fruits. To ensure formation of spurs on the laterals the central leader is cut back every year along with the strong erect laterals near the central leader. This leads to wide angled vigorous laterals for formation of spurs.

- Regulated System: Regulated pruning is practiced generally on apple cultivars growing on semi-dwarfing and vigorous rootstocks. Before planting, the central leader of the tree is cut back at 75 cm on which three well- placed primary branches are allowed to grow. In bearing trees, the growth of leader and strong laterals are encouraged by pruning weak and crowded branches.

- Renewal System: In vigorous cultivars instead of developing permanent spurs, the objective is to encourage continuous growth of new healthy shoots, spurs and branches every year. A part of the tree is pruned every year to produce fruits in the following year on the new shoot growth, while the unpruned parts produces fruit buds.

Thinning of Fruit

Thinning is one of the major techniques employed to regulate fruit quality. In apples, heavy bearing not only results in small-sized poor quality fruits but also sets in alternate bearing cycle. Judicious thinning done at the proper stage of fruit development can regulate cropping and improve fruit size and quality. Since manual thinning is cumbersome and expensive, chemical thinning is employed.

Chemical thinners should not be applied in very hot and dry conditions as it adversely affects the absorption. Spraying should be done thoroughly to cover the entire canopy. Sometimes chemical thinning follow calcium deficiency therefore adequate calcium nutrition should be supplemented after thinning. It is desirable to retain one fruit for every 40 leaves. This spaces the fruit at about 15-20 cm apart and there will be only one fruit per spur.

Manuring and Fertilization

Farmyard manure @ 10 kg./ year age of tree is applied along with other fertilizers. The fertilizer dose depends upon the fertility of soil and amount of organic manure applied to the crop. Generally, application of 350 g N, 175 g P_2O_5 and 350 g K_2O per plant per year in split doses is recommended for fully-grown bearing trees. On some trees deficiency of zinc, boron, manganese and calcium may be observed which is corrected with the application of appropriate chemicals through foliage spray.

Plant Protection

- Insect Pests: The insect pests mostly observed are San Jose Scale (*Quadraspidiotus perniciosus*), white scale (*Pseudoulacaspis* sp.), wooly apple aphid (*Eriosoma lanigerum*), blossom thrips (*Thrips rhopalantennalis*) etc. Planting of resistant rootstocks, suitable intercultural operations and spraying with chloropyriphos, fenitrothion, carbaryl etc. have been found to be effective in controlling the pests.

- Diseases: The main diseases reported are collar rot (*Phytophthora cactorum*), apple scab (*Venturia inaequalis*), sclerotius blight (*Sclerotium rolfsii*), crown gall (*Agrobacterium tumefaciens*), cankers, die-back diseases etc. Plants resistant to the diseases should be used for cultivation. The infected plant parts need to be destroyed. Application of copper oxychloride, carbendazim, mancozeb and other fungicides have been found to be effective in controlling the diseases.

- Disorders: In apple, there are three distinct fruit drops.

- Early drop resulting from unpollinated or unfertilized blossoms.

- June drop (due to moisture stress and fruit competition).

- Pre-harvest drop. Pre-harvest drop can be controlled by spraying NAA @ 10 ppm. (1 ml. of Planofix dissolved in 4.5 l. of water) about a week before the expected drop.

Harvesting

Normally the apples are ready for harvest from September-October except in the Nilgiris where the season is from April to July. The fruits mature within 130-150 days after the full bloom stage depending upon the variety grown. The ripening of fruits is associated with the change in colour, texture, quality and the development of the characteristic flavour. The fruits at the time of harvest should be uniform, firm and crisp. The colour of the skin at maturity ranges from yellow-red depending on the variety. However, the optimum time of harvest depends on fruit quality and intended period of storage. Due to the introduction of dwarf rootstock hand picking is recommended as it reduces bruising due to fruit fall during mechanical harvesting. Yield: The apple tree starts bearing from 4 year onwards. Depending on variety and season, a well-managed apple orchard yields on an average 10-20 kg/tree/year.

Strawberry

Strawberry (*Fragaria vesca*) is an important fruit crop and its commercial production is possible in temperate and sub-tropical areas of the country. But varieties are available which can be cultivated in the subtropical climate. it is generally cultivated in the hills. The strawberry is the most widely adapted of the small fruits. Strawberries are grown throughout Europe, in every state of the United States, as well as in Canada and South America. The wide variation in climates within these regions and the wide adaptation of the strawberry plant permit harvesting and marketing, the fruit during the greater part of the year.

Economic Importance of strawberry production: Strawberry is rich in Vitamin C and iron. Some varieties viz. Olympus, Hood & Shuksan having high avor and bright red color are suitable for ice-cream making. Other varieties like Midway, Midland, Cardinal, Hood, Redchief, and Beauty are best for processing.

Suitable Climate and Soil for Strawberry Plantation

- Strawberry grown best in a temperate climate. It is a short day plant, which requires exposure to about 10 days of less than 8 hours of sunshine for the initiation of owering. In winter, the plants do not make any growth and remain dormant. The exposure to low temperature during this period helps in breaking the dormancy of the plant. In spring when the days become longer and the temperature rises. The plants continue to grow and start owering. The varieties grown in milder subtropical climate do not require chilling and continue to make some growth during winter.

- From the standpoint of response to the length of the light period, strawberries are placed in two groups: (1) varieties that develop ower buds during both long and short light periods, the overbearing varieties and (2) varieties that develop ower buds during the short light periods only, most commercial varieties.

- Strawberry requires a well-drained medium loam soil, rich in organic matter. The soil should be slightly acidic with a pH from 5.7 to 6.5. At higher pH (acidic) root formation is very poor. The presence of high calcium in the soil causes yellowing of the leaves. In light soils and in those rich in organic matter, runner formation is better. Strawberry should not be cultivated in the same land for a number of years. It is preferable to plant it in the green manured eld. Alkaline soils and soils infected with nematodes should not be considered for strawberry farming and must be avoided.

Varieties of Strawberries

A large number of varieties are available. For the hilly areas, varieties Royal Sovereign, Srinagar and Dilpasand are suitable. Some of the introductions from California, such as Torrey, Toiga, and Solana may prove even more successful. The variety found successful in Bangalore has been named Bangalore and which has performed well at Mahabaleshwar also. For the north Indian plains, Pusa Early Dwarf which has dwarf plants, large rm wedge-shaped fruits, has been recommended. Another variety with rich aroma but softer fruits is Katrain Sweet. Some of the varieties found successful in warmer parts of the U.S.A. are Premier Florida-90, Missionary, Blackmore, Klonmore & Klondike. Some of these may prove successful for cultivation in Indian plains.

Propagation of Strawberry Plants

Propagation is done by means of runners that are formed after the blooming season. The plants may be allowed to set as many runners as possible but not allowed to set any fruits. All the plants with a good root system should be utilized to set a new plantation. Given the best attention and care, a single plant usually produces 12 to 18 runners.

Best Planting Season for Strawberries

The ideal time of planting runners or crowns in hilly areas is September-October. If the planting is done too early, plants lack vigor and result in low yield and quality of fruits. If planted very late, runners develop in March and crops are light.

Runners are uprooted from the nursery, made into bundles and planted in the eld. These can be kept in cold storage before transplanting. The soil should be frequently irrigated to reduce water stress in the leaf. Defoliation suppresses the plant growth, delays fruiting and reduces yield & quality.

Varieties of Strawberries

- In strawberry farming, the land for strawberry planting should be thoroughly prepared by deep ploughing followed by harrowing. Liberal quantities of organic manure should be incorporated in the soil before planting. Strawberry can be planted on at beds, in the form of hill rows or matted rows, or it can be planted on raised beds. In irrigated areas, plantings on ridges are advised. In Mahabaleshwar, the usual practice is to plant on raised beds 4 x 3 meters or 4 × 4 meters.

- In strawberry farming, planting distance varies according to variety & type of land. A spacing of 30 cm. x 60 cm. is usually followed. In the model scheme, a spacing of 30 cm. x 30 cm. with a population of 22,000 plants per acre has been considered which was commonly observed in areas covered during a eld study. from row to row. In the hills, Transplanting is done in March-April, September-October, but in the plains, the months of January-February may be utilized for this purpose. At Mahabaleshwar, Tamilanadu, normally strawberry is planted during November December.

- The strawberry plants should be set in the soil with their roots going straight down. The soil around the plant should be rmly packed to exclude air. The growing point of the plant should be just above the soil surface. During planting, the plants should not be allowed to dry out and should be irrigated immediately after planting.

Plant Care in Strawberry Production

The roots of strawberry plants spread out close to the surface. Therefore, the soil should be well supplied with moisture, and hoeing should be done lightly and young plantation is kept weed free.

Horticultural Practices of the Strawberry Crop

In cold climatic conditions, the soil is covered with mulch in winter to protect the roots from cold injury. The mulch keeps the fruits free from soil, reduces decay of fruits, conserves soil moisture, lowers soil temperature in hot weather, protects owers from frost in mild climates and protects plants from freezing injury in cold climates. Several kinds of mulches are used, but the commonest one is straw mulch. The name strawberry has been derived from this fact. Black alkathine mulch is also used to cover the soil. It saves irrigation water, prevents the growth of weeds and keeps the soil temperature high.

Irrigation Requirement of Strawberry Plantation

Strawberry being a shallow-rooted plant requires more frequent but less amount of water in each irrigation. Excessive irrigation results in the growth of leaves and stolons at the expense of fruits & owers and also increase the incidence of Botrytis rot. Since strawberry is relatively shallow-rooted, it is susceptible to conditions of drought. Planting early in autumn allows the plants to make good vegetative growth before the onset of winter.

However, in this case, it is necessary to ensure that newly planted runners are irrigated frequently after planting, otherwise, the mortality of the plants becomes high. During September and October, irrigation should be given twice a week if there is no rain. It may be reduced to weekly intervals during November. In December and January, irrigation may be given once every fortnight. When fruiting starts, the irrigation frequency may again be increased. At this stage, frequent irrigation gives larger fruits.

Manures and Fertilizers Applications for Strawberry Plants

A fertilizer dose of 25-50 tonnes farmyard manure, 75-100 kg. N, 40-120 kg. P_2O_5, 40-80 kg. K_2O/ ha. may be applied according to soil type and variety planted.

Intercultural Operations of the Strawberry Crop

The eld is kept weed free during the rst season by harrowing and plowing, applying herbicides or plastic sheet. Inter-cultural practices are continued until the straw mulch is applied.

Plant Protection Measures in the Strawberry Orchard

Pests and Diseases of Strawberry

- Red spider mites and cutworms are important pests of strawberry. The mites can be controlled with 0.05 percent Monocrotophos + 0.25 percent wettable sulfur. The cutworms must be controlled by dusting the soil before planting with 5 percent chlordane or Heptachlor dust at the rate of 50 kg per hectare and mixing it thoroughly in the soil by the cultivator.

- The two commonest diseases of strawberry are red stele, caused by the fungus Phytophthora fragariae and black root rot. The remedy for the former lies by raising resistant varieties like stelemaster and for the latter to maintain the vigor of the plants and rotate strawberry with other crops like legume vegetables (beans, peas, etc). Strawberry also suers from virus diseases known as a yellow edge, crinkle, and dwarf. Raising of a strawberry nursery in the hills helps to check these. Strawberry also throws some chlorotic plants, which result from genetic segregation. These should not be confused with virus aected plants and should be rogued out.

Harvesting and Yield of Strawberry Farming

Ripened Strawberry

- Strawberries are generally harvested when half to three-fourths of the skin develops color. Depending on the weather conditions, picking is usually done on every second or third day usually in the morning hours. Strawberries are harvested in small trays or baskets. They should be kept in a shady place to avoid damage due to excessive heat in the open eld. The fruit ripens from late February to April in the plains and during May and June at high elevations like Mahabaleshwar, Nainital, and Kashmir. For the local market, the fruit should be harvested when fully the ripe stage, but for transport to distant markets, it should be harvested when still rm and before color has developed fully all over the fruit.

- Harvesting should be done preferably daily. Since fruit is highly perishable, it is packed in at shallow containers of various types (cardboard, bamboo, paper trays, etc.) with one or two layers of fruits. Harvesting should be done early in the morning in dry conditions. Washing the fruit bruises it and spoils its lustre.

- The yield of strawberry farming varies according to season and locality. A yield of 20 to 25 tons per hectare is excellent, though yields up to 50 tons per hectare have been reported under ideal conditions.

- Plants start bearing in the second year. An average yield of 45-100 quintals/ha. is obtained from a strawberry orchard. However, an average yield of 175-300 q./ha. may be taken from a well-managed strawberry orchard.

Post-harvesting Care and Marketing

In your strawberry farming, care should be taken as strawberries are highly perishable and hence a great deal of care in harvesting and handling as well as its marketing also requires to be organized very carefully.

- Grading: Fruits are graded on the basis of their weight, size, and color.

- Storage: Fruits can be stored in cold storage at 32 °C up to10 days. For distant marketing, strawberries should be pre-cooled at 4 °C within 2 hrs. of harvesting and kept at the same temperature. After pre-cooling, they are shipped in refrigerated vans.

- Packing: Packing is done according to the grades for long distance markets. Fruits of good quality are packed in perforated cardboard cartons with paper cuttings as cushioning material. Fruits of lower grades are packed in baskets.

- Transportation: Road transport by trucks/lorries is the most convenient mode of transport due to the easy approach from orchards to the market.

- Marketing: Majority of the growers sell their produce either through trade agents at village level or commission agents at the market.

Post-harvesting Strawberries.

References

- Guide-capsicum-production-technologies, agripedia: krishijagran.com, Retrieved 10, July 2020

- Crop-guide-growing-cucumbers, cucumber-fertilizer: haifa-group.com, Retrieved 19, January 2020

- Pumpkin, crops, vegetables: celkau.in, Retrieved 14, August 2020

- Guava-fruit-farming: agrifarming.in, Retrieved 27, April 2020

- Banana-farming-information-guide: agrifarming.in, Retrieved 11, February 2020

- Apple, fruits, package-of-practices, crop-production, agriculture: vikaspedia.in, Retrieved 17, May 2020

- Strawberry-farming: agrifarming.in, Retrieved 23, March 2020

Chapter 3

An Overview of Orchard

An orchard is a plantation of trees or shrubs that is maintained for the production of food. Orchards comprise fruit-producing and nut-producing trees which are generally grown for commercial purposes. This chapter has been carefully written to provide an easy understanding of the varied facets of orchards.

A lemon orchard in the Upper Galilee in Israel.

An orchard is an intentional planting of trees or shrubs that is maintained for food production. Orchards comprise fruit- or nut-producing trees which are generally grown for commercial production. Orchards are also sometimes a feature of large gardens, where they serve an aesthetic as well as a productive purpose. A fruit garden is generally synonymous with an orchard, although it is set on a smaller non-commercial scale and may emphasize berry shrubs in preference to fruit trees. Most temperate-zone orchards are laid out in a regular grid, with a grazed or mown grass or bare soil base that makes maintenance and fruit gathering easy.

Orchards are sometimes concentrated near bodies of water, where climatic extremes are moderated and blossom time is retarded until frost danger is past.

Layout

An orchard's layout is the technique of planting the crops in a proper system. There are different methods of planting and thus different layouts. Some of these layout types include:

1. Square method

2. Rectangular method

3. Quincunx method

4. Triangular method

5. Hexagonal method

6. Contour (or Terracing) method

For different varieties, these systems may vary to some extent.

Orchards by Region

Apple orchards in Azwell, Washington surrounding a community of pickers' cabins.

Sour cherry orchard on Lake Erie shoreline (Leamington, Ontario).

The most extensive orchards in the United States are apple and orange orchards, although citrus orchards are more commonly called groves. The most extensive apple orchard area is in eastern Washington state, with a lesser but significant apple orchard area in most of Upstate New York. Extensive orange orchards are found in Florida and southern California,where they are more widely known as 'groves'. In eastern North America, many orchards are along the shores of Lake Michigan (such as the Fruit Ridge Region), Lake Erie, and Lake Ontario.

In Canada, apple and other fruit orchards are widespread on the Niagara Peninsula, south of Lake Ontario. This region is known as Canada Fruitbelt and, in addition to large-scale commercial fruit marketing, it encourages "pick-your-own" activities in the harvest season.

Murcia is a major orchard area (or la huerta) in Europe, with citrus crops. New Zealand, China, Argentina and Chile also have extensive apple orchards.

Tenbury Wells in Worcestershire has been called The Town in the Orchard, since the 19th century, because it was surrounded by extensive orchards. Today, this heritage is celebrated through an annual Applefest.

Central Europe

Streuobstwiese (pl. Streuobstwiesen) is a German word that means a meadow with scattered fruit trees or fruit trees that are planted in a field. Streuobstwiese, or a meadow orchard, is a traditional landscape in the temperate, maritime climate of continental Western Europe. In the 19th and early 20th centuries, Streuobstwiesen were a kind of a rural community orchard that were intended for productive cultivation of stone fruit. In recent years, ecologists have successfully lobbied for state subsidies to valuable habitats, biodiversity and natural landscapes, which are also used to preserve old meadow orchards. Both conventional and meadow orchards provide a suitable habitat for many animal species that live in a cultured landscape. A notable example is the hoopoe that nests in tree hollows of old fruit trees and, in the absence of alternative nesting sites, is threatened in many parts of Europe, because of the destruction of old orchards.

Historical Orchards

Old growth apple orchard in Ottawa, Canada.

- Orchard House in Concord, Massachusetts was the residence of American celebrated writer Louisa May Alcott.

- Fruita, Utah part of Capitol Reef National Park has Mormon pioneer orchards maintained by the United States National Park Service.

Orchard Conservation in the UK

- Natural England, through its Countryside Stewardship Scheme, Environmental

Stewardship and Environmentally Sensitive Areas Scheme, gives grant aid and advice for the maintenance, enhancement or re-creation of historical orchards.

- The 'Orchard Link' organisation provides advice on how to manage and restore the county of Devon's orchards, as well as enabling the local community to use the local orchard produce. An organisation called 'Orchards Live' carries out similar work in North Devon.

- People's Trust for Endangered Species (PTES) has mapped every traditional orchard within England and Wales and manages the national inventory for this habitat.

- The UK Biodiversity Partnership lists traditional orchards and a priority UK Biodiversity Action Plan habitat.

- The Wiltshire Traditional Orchards Project maps, conserves and restores traditional orchards within Wiltshire, England.

Fruit

Culinary fruits.

Several culinary fruits.

In botany, a fruit is the seed-bearing structure in flowering plants (also known as angiosperms) formed from the ovary after flowering.

Mixed fruit.

The Medici citrus collection by Bartolomeo Bimbi, 1715.

Fruits are the means by which angiosperms disseminate seeds. Edible fruits, in particular, have propagated with the movements of humans and animals in a symbiotic relationship as a means for seed dispersal and nutrition; in fact, humans and many animals have become dependent on fruits as a source of food. Accordingly, fruits account for a substantial fraction of the world's agricultural output, and some (such as the apple and the pomegranate) have acquired extensive cultural and symbolic meanings.

In common language usage, "fruit" normally means the fleshy seed-associated structures of a plant that are sweet or sour, and edible in the raw state, such as apples, bananas, grapes, lemons, oranges, and strawberries. On the other hand, in botanical usage, "fruit" includes many structures that are not commonly called "fruits", such as bean pods, corn kernels, tomatoes, and wheat grains. The section of a fungus that produces spores is also called a fruiting body.

Botanic Fruit and Culinary Fruit

Many common terms for seeds and fruit do not correspond to the botanical classifications. In culinary terminology, a fruit is usually any sweet-tasting plant part, especially a botanical fruit; a nut is any hard, oily, and shelled plant product; and a vegetable is any savory or less sweet plant product. However, in botany, a fruit is the ripened ovary or carpel that contains seeds, a nut is a type of fruit and not a seed, and a seed is a ripened ovule.

Venn diagram representing the relationship between (culinary) vegetables and botanical fruits.

Examples of culinary "vegetables" and nuts that are botanically fruit include corn, cucurbits (e.g., cucumber, pumpkin, and squash), eggplant, legumes (beans, peanuts, and peas), sweet pepper, and tomato. In addition, some spices, such as allspice and chili pepper, are fruits, botanically speaking. In contrast, rhubarb is often referred to as a fruit, because it is used to make sweet desserts such as pies, though only the petiole (leaf stalk) of the rhubarb plant is edible, and edible gymnosperm seeds are often given fruit names, e.g., ginkgo nuts and pine nuts.

Botanically, a cereal grain, such as corn, rice, or wheat, is also a kind of fruit, termed a caryopsis. However, the fruit wall is very thin and is fused to the seed coat, so almost all of the edible grain is actually a seed.

Fruit Structure

The outer, often edible layer, is the pericarp, formed from the ovary and surrounding the seeds, although in some species other tissues contribute to or form the edible portion. The pericarp may be described in three layers from outer to inner, the epicarp, mesocarp and endocarp.

Fruit that bears a prominent pointed terminal projection is said to be beaked.

Fruit Development

A fruit results from maturation of one or more flowers, and the gynoecium of the flower(s) forms all or part of the fruit.

Inside the ovary/ovaries are one or more ovules where the megagametophyte contains the egg cell. After double fertilization, these ovules will become seeds. The ovules are fertilized in a process that starts with pollination, which involves the movement of pollen from the stamens to the stigma of flowers. After pollination, a tube grows from the pollen through the stigma into the ovary to the ovule and two sperm are transferred from the pollen to the megagametophyte. Within the megagametophyte one of the two sperm unites with the egg, forming a zygote, and the second sperm enters the central cell forming the endosperm mother cell, which completes the double fertilization process. Later the zygote will give rise to the embryo of the seed, and the endosperm mother cell will give rise to endosperm, a nutritive tissue used by the embryo.

The development sequence of a typical drupe, the nectarine (Prunus persica) over a 7.5 month period, from bud formation in early winter to fruit ripening in midsummer.

As the ovules develop into seeds, the ovary begins to ripen and the ovary wall, the pericarp, may become fleshy (as in berries or drupes), or form a hard outer covering (as in nuts). In some multiseeded fruits, the extent to which the flesh develops is proportional to the number of fertilized ovules. The pericarp is often differentiated into two or three distinct layers called the exocarp (outer layer, also called epicarp), mesocarp (middle layer), and endocarp (inner layer). In some fruits, especially simple fruits derived from an inferior ovary, other parts of the flower (such as the floral tube, including the petals, sepals, and stamens), fuse with the ovary and ripen with it. In other cases, the sepals, petals and/or stamens and style of the flower fall off. When such other floral parts are a significant part of the fruit, it is called an accessory fruit. Since other parts of the flower may contribute to the structure of the fruit, it is important to study flower structure to understand how a particular fruit forms.

There are three general modes of fruit development:

- Apocarpous fruits develop from a single flower having one or more separate carpels, and they are the simplest fruits.

- Syncarpous fruits develop from a single gynoecium having two or more carpels fused together.

- Multiple fruits form from many different flowers.

Plant scientists have grouped fruits into three main groups, simple fruits, aggregate fruits, and composite or multiple fruits. The groupings are not evolutionarily relevant, since many diverse plant taxa may be in the same group, but reflect how the flower organs are arranged and how the fruits develop.

Simple Fruit

Epigynous berries are simple fleshy fruit. Clockwise from top right: cranberries, lingonberries, blueberries, red huckleberries.

Simple fruits can be either dry or fleshy, and result from the ripening of a simple or compound ovary in a flower with only one pistil. Dry fruits may be either dehiscent (they open to discharge seeds), or indehiscent (they do not open to discharge seeds). Types of dry, simple fruits, and examples of each, include:

- Achene – Most commonly seen in aggregate fruits (e.g., strawberry).

- Capsule – (e.g., Brazil nut).

- Caryopsis – (e.g., wheat).

- Cypsela – an achene-like fruit derived from the individual florets in a capitulum (e.g., dandelion).

- Fibrous drupe – (e.g., coconut, walnut).

- Follicle – is formed from a single carpel, opens by one suture (e.g., milkweed), commonly seen in aggregate fruits (e.g., magnolia).

- Legume – (e.g., bean, pea, peanut).

- Loment – a type of indehiscent legume.

- Nut – (e.g., beech, hazelnut, oak acorn).

- Samara – (e.g., ash, elm, maple key).

- Schizocarp – (e.g., carrot seed).

- Silique – (e.g., radish seed).

- Silicle – (e.g., shepherd's purse).

- Utricle – (e.g., beet).

Lilium unripe capsule fruit.

Fruits in which part or all of the pericarp (fruit wall) is fleshy at maturity are simple fleshy fruits. Types of simple, fleshy, fruits (with examples) include:

- Berry – (e.g., cranberry, gooseberry, redcurrant, tomato).

- Stone fruit or drupe (e.g., apricot, cherry, olive, peach, plum).

Dewberry flowers. Note the multiple pistils, each of which will produce a drupelet. Each flower will become a blackberry-like aggregate fruit.

An aggregate fruit, or etaerio, develops from a single flower with numerous simple pistils.

- Magnolia and peony, collection of follicles developing from one flower.

- Sweet gum, collection of capsules.

- Sycamore, collection of achenes.

- Teasel, collection of cypsellas

- Tuliptree, collection of samaras.

The pome fruits of the family Rosaceae, (including apples, pears, rosehips, and saskatoon berry) are a syncarpous fleshy fruit, a simple fruit, developing from a half-inferior ovary.

Schizocarp fruits form from a syncarpous ovary and do not really dehisce, but rather split into segments with one or more seeds; they include a number of different forms from a wide range of families. Carrot seed is an example.

Aggregate Fruit

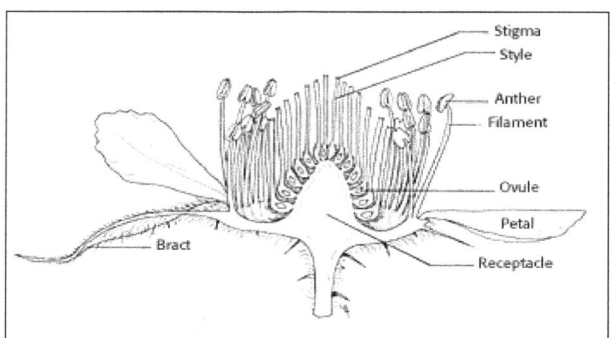

Detail of raspberry flower.

Aggregate fruits form from single flowers that have multiple carpels which are not joined together, i.e. each pistil contains one carpel. Each pistil forms a fruitlet, and collectively the fruitlets are called an etaerio. Four types of aggregate fruits include etaerios of achenes, follicles, drupelets, and berries. Ranunculaceae species, including Clematis and Ranunculus have an etaerio of achenes, Calotropis has an etaerio of follicles, and Rubus species like raspberry, have an etaerio of drupelets. Annona have an etaerio of berries.

The raspberry, whose pistils are termed drupelets because each is like a small drupe attached to the receptacle. In some bramble fruits (such as blackberry) the receptacle is elongated and part of the ripe fruit, making the blackberry an aggregate-accessory fruit. The strawberry is also an aggregate-accessory fruit, only one in which the seeds are contained in achenes. In all these examples, the fruit develops from a single flower with numerous pistils.

Multiple Fruits

A multiple fruit is one formed from a cluster of flowers (called an inflorescence). Each flower produces a fruit, but these mature into a single mass. Examples are the pineapple, fig, mulberry, osage-orange, and breadfruit.

In the photograph on the right, stages of flowering and fruit development in the noni or Indian mulberry (Morinda citrifolia) can be observed on a single branch. First an inflorescence of white flowers called a head is produced. After fertilization, each flower develops into a drupe, and as the drupes expand, they become connate (merge) into a multiple fleshy fruit called a syncarp.

In some plants, such as this noni, flowers are produced regularly along the stem and it is possible to see together examples of flowering, fruit development, and fruit ripening.

Berries

Berries are another type of fleshy fruit; they are simple fruit created from a single ovary. The ovary may be compound, with several carpels. Types include (examples follow in the table below):

- Pepo – berries whose skin is hardened, cucurbits.

- Hesperidium – berries with a rind and a juicy interior, like most citrus fruit.

Accessory Fruit

The fruit of a pineapple includes tissue from the sepals as well as the pistils of many flowers. It is an accessory fruit and a multiple fruit.

Some or all of the edible part of accessory fruit is not generated by the ovary. Accessory fruit can be simple, aggregate, or multiple, i.e., they can include one or more pistils and other parts from the same flower, or the pistils and other parts of many flowers.

Table of fruit examples.

Types of fleshy fruits					
True berry	Pepo	Hesperidium	Aggregate fruit	Multiple fruit	Accessory fruit
Blackcurrant, Blueberry, Chili pepper, Cranberry, Eggplant, Gooseberry, Grape, Guava, Kiwifruit, Lucuma, Pomegranate, Redcurrant, Tomato	Cucumber, Gourd, Melon, Pumpkin	Grapefruit, Lemon, Lime, Orange	Blackberry, Boysenberry, Raspberry	Fig, Hedge apple, Mulberry, Pineapple	Apple, Pineapple, Rose hip, Stone fruit, Strawberry

Seedless Fruits

An arrangement of fruits commonly thought of as vegetables, including tomatoes and various squash.

Seedlessness is an important feature of some fruits of commerce. Commercial cultivars of bananas and pineapples are examples of seedless fruits. Some cultivars of citrus fruits (especially grapefruit, mandarin oranges, navel oranges), satsumas, table grapes, and watermelons are valued for their seedlessness. In some species, seedlessness is the result of parthenocarpy, where fruits set without fertilization. Parthenocarpic fruit set may or may not require pollination, but most seedless citrus fruits require a stimulus from pollination to produce fruit.

Seedless bananas and grapes are triploids, and seedlessness results from the abortion of the embryonic plant that is produced by fertilization, a phenomenon known as stenospermocarpy, which requires normal pollination and fertilization.

Seed Dissemination

Variations in fruit structures largely depend on their seeds' mode of dispersal. This dispersal can be achieved by animals, explosive dehiscence, water, or wind.

Some fruits have coats covered with spikes or hooked burrs, either to prevent themselves from being eaten by animals, or to stick to the feathers, hairs, or legs of animals, using them as dispersal agents. Examples include cocklebur and unicorn plant.

Grapes and Mangoes.

The sweet flesh of many fruits is "deliberately" appealing to animals, so that the seeds held within are eaten and "unwittingly" carried away and deposited (i.e., defecated) at a distance from the parent. Likewise, the nutritious, oily kernels of nuts are appealing to rodents (such as squirrels), which hoard them in the soil to avoid starving during the winter, thus giving those seeds that remain uneaten the chance to germinate and grow into a new plant away from their parent.

Other fruits are elongated and flattened out naturally, and so become thin, like wings or helicopter blades, e.g., elm, maple, and tuliptree. This is an evolutionary mechanism to increase dispersal distance away from the parent, via wind. Other wind-dispersed fruit have tiny "parachutes", e.g., dandelion, milkweed, salsify.

Coconut fruits can float thousands of miles in the ocean to spread seeds. Some other fruits that can disperse via water are nipa palm and screw pine.

Some fruits fling seeds substantial distances (up to 100 m in sandbox tree) via explosive dehiscence or other mechanisms, e.g., impatiens and squirting cucumber.

Uses

Nectarines are one of many fruits that can be easily stewed.

Many hundreds of fruits, including fleshy fruits (like apple, kiwifruit, mango, peach, pear, and watermelon) are commercially valuable as human food, eaten both fresh and as jams, marmalade and other preserves. Fruits are also used in manufactured foods (e.g., cakes, cookies, ice cream, muffins, or yogurt) or beverages, such as fruit juices (e.g., apple juice, grape juice, or orange juice) or alcoholic beverages (e.g., brandy, fruit beer, or wine), Fruits are also used for gift giving, e.g., in the form of Fruit Baskets and Fruit Bouquets.

Oranges, bananas, pears, apples, and a watermelon.

Many "vegetables" in culinary parlance are botanical fruits, including bell pepper, cucumber, eggplant, green bean, okra, pumpkin, squash, tomato, and zucchini. Olive fruit is pressed for olive oil. Spices like allspice, black pepper, paprika, and vanilla are derived from berries.

Nutritional Value

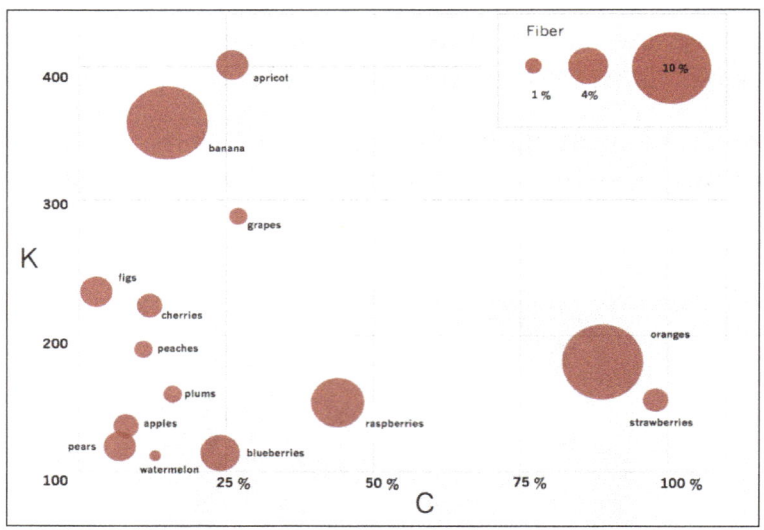

Each point refers to a 100 g serving of the fresh fruit, the daily recommended allowance of vitamin C is on the X axis and mg of Potassium (K) on the Y (offset by 100 mg which every fruit has) and the size of the disk represents amount of fiber (key in upper right). Watermelon, which has almost no fiber, and low levels of vitamin C and potassium, comes in last place.

Fresh fruits are generally high in fiber, vitamin C, and water.

Regular consumption of fruit is generally associated with reduced risks of several diseases and functional declines associated with aging.

Nonfood Uses

Because fruits have been such a major part of the human diet, various cultures have developed many different uses for fruits they do not depend on for food. For example:

- Bayberry fruits provide a wax often used to make candles.

- Many dry fruits are used as decorations or in dried flower arrangements (e.g., annual honesty, cotoneaster, lotus, milkweed, unicorn plant, and wheat). Ornamental trees and shrubs are often cultivated for their colorful fruits, including beautyberry, cotoneaster, holly, pyracantha, skimmia, and viburnum.

- Fruits of opium poppy are the source of opium, which contains the drugs codeine and morphine, as well as the biologically inactive chemical thebaine from which the drug oxycodone is synthesized.

- Osage orange fruits are used to repel cockroaches.

- Many fruits provide natural dyes (e.g., cherry, mulberry, sumac, and walnut).

- Dried gourds are used as bird houses, cups, decorations, dishes, musical instruments, and water jugs.

- Pumpkins are carved into Jack-o'-lanterns for Halloween.

- The spiny fruit of burdock or cocklebur inspired the invention of Velcro.

- Coir fiber from coconut shells is used for brushes, doormats, floor tiles, insulation, mattresses, sacking, and as a growing medium for container plants. The shell of the coconut fruit is used to make bird houses, bowls, cups, musical instruments, and souvenir heads.

- Fruit is often a subject of still life paintings.

Safety

For food safety, the CDC recommends proper fruit handling and preparation to reduce the risk of food contamination and foodborne illness. Fresh fruits and vegetables should be carefully selected; at the store, they should not be damaged or bruised; and pre-cut pieces should be refrigerated or surrounded by ice.

All fruits and vegetables should be rinsed before eating. This recommendation also applies to produce with rinds or skins that are not eaten. It should be done just before preparing or eating to avoid premature spoilage.

Fruits and vegetables should be kept separate from raw foods like meat, poultry, and seafood, as well as from utensils that have come in contact with raw foods. Fruits and vegetables that are not going to be cooked should be thrown away if they have touched raw meat, poultry, seafood, or eggs.

All cut, peeled, or cooked fruits and vegetables should be refrigerated within two hours. After a certain time, harmful bacteria may grow on them and increase the risk of food-borne illness.

Allergies

Fruit allergies make up about 10 percent of all food related allergies.

Storage

All fruits benefit from proper post harvest care, and in many fruits, the plant hormone ethylene causes ripening. Therefore, maintaining most fruits in an efficient cold chain is optimal for post harvest storage, with the aim of extending and ensuring shelf life.

Vegetable

Vegetables in a market in the Philippines.

In everyday usage, a vegetable is any part of a plant that is consumed by humans as food as part of a savory meal. The term vegetable is somewhat arbitrary, and largely defined through culinary and cultural tradition. It normally excludes other food derived from plants such as fruits, nuts, and cereal grains, but includes seeds such as pulses. The original meaning of the word vegetable, still used in biology, was to describe all types of plant, as in the terms "vegetable kingdom" and "vegetable matter".

Originally, vegetables were collected from the wild by hunter-gatherers and entered cultivation in several parts of the world, probably during the period 10,000 BC to 7,000 BC, when a new agricultural way of life developed. At first, plants which grew locally would have been cultivated, but as time went on, trade brought exotic crops from elsewhere to add to domestic types. Nowadays, most vegetables are grown all over the world as climate permits, and crops may be cultivated in protected environments in less suitable locations. China is the largest producer of vegetables and global trade in agricultural products allows consumers to purchase vegetables grown in faraway countries. The scale of production varies from subsistence farmers supplying the needs of their family for food, to agribusinesses with vast acreages of single-product crops. Depending on the type of vegetable concerned, harvesting the crop is followed by grading, storing, processing, and marketing.

Vegetables can be eaten either raw or cooked and play an important role in human nutrition, being mostly low in fat and carbohydrates, but high in vitamins, minerals and fiber. Many nutritionists encourage people to consume plenty of fruit and vegetables, five or more portions a day often being recommended.

Etymology

Domestic vegetable garden in the United Kingdom.

The word vegetable was first recorded in English in the early 15th century. It comes from Old French, and was originally applied to all plants; the word is still used in this sense in biological contexts. It derives from Medieval Latin vegetabilis "growing, flourishing" (i.e. of a plant), a semantic change from a Late Latin meaning "to be enlivening, quickening".

The meaning of "vegetable" as a "plant grown for food" was not established until the 18th century. In 1767, the word was specifically used to mean a "plant cultivated for food, an edible herb or root". The year 1955 noted the first use of the shortened, slang term "veggie".

As an adjective, the word vegetable is used in scientific and technical contexts with a different and much broader meaning, namely of "related to plants" in general, edible or not — as in vegetable matter, vegetable kingdom, vegetable origin, etc.

Terminology

The exact definition of "vegetable" may vary simply because of the many parts of a plant consumed as food worldwide – roots, tubers, bulbs, corms, stems, leaf stems, leaf sheaths, leaves, buds, flowers, fruits, and seeds. The broadest definition is the word's use adjectivally to mean "matter of plant origin" to distinguish it from "animal", meaning "matter of animal origin". More specifically, a vegetable may be defined as "any plant, part of which is used for food", a secondary meaning then being "the edible part of such a plant". A more precise definition is "any plant part consumed for food that is not a fruit or seed, but including mature fruits that are eaten as part of a main meal". Falling outside these definitions are edible fungi (such as mushrooms) and edible seaweed which, although not parts of plants, are often treated as vegetables.

A Venn diagram shows the overlap in the terminology of "vegetables" in a culinary sense and "fruits" in the botanical sense.

In everyday language, the words "fruit" and "vegetable" are mutually exclusive. "Fruit" has a precise botanical meaning, being a part that developed from the ovary of a flowering plant. This is considerably different from the word's culinary meaning. While peaches, plums, and oranges are "fruit" in both senses, many items commonly called "vegetables", such as eggplants, bell peppers, and tomatoes, are botanically fruits. The question of whether the tomato is a fruit or a vegetable found its way into the United States Supreme Court in 1893. The court ruled unanimously in Nix v. Hedden that a tomato is correctly identified as, and thus taxed as, a vegetable, for the purposes of the Tariff of 1883 on imported produce. The court did acknowledge, however, that, botanically speaking, a tomato is a fruit.

History

Before the advent of agriculture, humans were hunter-gatherers. They foraged for edible fruit, nuts, stems, leaves, corms, and tubers, scavenged for dead animals and hunted living ones for food. Forest gardening in a tropical jungle clearing is thought to be the first example of agriculture; useful plant species were identified and encouraged to grow while undesirable species were removed. Plant breeding through the selection of strains with desirable traits such as large fruit and vigorous growth soon followed. While the first evidence for the domestication of grasses such as wheat and barley has been found in the Fertile Crescent in the Middle East, it is likely that various peoples around the world started growing crops in the period 10,000 BC to 7,000 BC. Subsistence agriculture continues to this day, with many rural farmers in Africa, Asia, South

America, and elsewhere using their plots of land to produce enough food for their families, while any surplus produce is used for exchange for other goods.

Throughout recorded history, the rich have been able to afford a varied diet including meat, vegetables and fruit, but for poor people, meat was a luxury and the food they ate was very dull, typically comprising mainly some staple product made from rice, rye, barley, wheat, millet or maize. The addition of vegetable matter provided some variety to the diet. The staple diet of the Aztecs in Central America was maize and they cultivated tomatoes, avocados, beans, peppers, pumpkins, squashes, peanuts, and amaranth seeds to supplement their tortillas and porridge. In Peru, the Incas subsisted on maize in the lowlands and potatoes at higher altitudes. They also used seeds from quinoa, supplementing their diet with peppers, tomatoes, and avocados.

In Ancient China, rice was the staple crop in the south and wheat in the north, the latter made into dumplings, noodles, and pancakes. Vegetables used to accompany these included yams, soybeans, broad beans, turnips, spring onions, and garlic. The diet of the ancient Egyptians was based on bread, often contaminated with sand which wore away their teeth. Meat was a luxury but fish was more plentiful. These were accompanied by a range of vegetables including marrows, broad beans, lentils, onions, leeks, garlic, radishes, and lettuces.

The mainstay of the Ancient Greek diet was bread, and this was accompanied by goat's cheese, olives, figs, fish, and occasionally meat. The vegetables grown included onions, garlic, cabbages, melons, and lentils. In Ancient Rome, a thick porridge was made of emmer wheat or beans, accompanied by green vegetables but little meat, and fish was not esteemed. The Romans grew broad beans, peas, onions and turnips and ate the leaves of beets rather than their roots.

Some Common Vegetables

Some common vegetables					
Image	Description	Parts used	Origin	Cultivars	World production ($\times 10^6$ tons, 2012)
	cabbage Brassica oleracea	leaves, axillary buds, stems, flower-heads	Europe	cabbage, red cabbage, Savoy cabbage, kale, Brussels sprouts, kohlrabi, cauliflower, broccoli, Chinese broccoli	70.1

	turnip Brassica rapa	tubers, leaves	Asia	turnip, rutabaga, Chinese cabbage, napa cabbage, bok choy, collard greens	
	radish Raphanus sativus	roots, leaves, seed pods, seed oil, sprouting	Southeastern Asia	radish, daikon, seedpod varieties	
	carrot Daucus carota	root tubers	Persia	carrot	36.9
	parsnip Pastinaca sativa	Root tubers	Eurasia	parsnip	
	beetroot Beta vulgaris	tubers, leaves	Europe, Near East, and India	beetroot, sea beet, Swiss chard, sugar beet	
	lettuce Lactuca sativa	leaves, stems, seed oil	Egypt	lettuce, celtuce	24.9
	beans Phaseolus vulgaris Phaseolus coccineus Phaseolus lunatus	pods, seeds	Central and South America	green bean, French bean, runner bean, haricot bean, Lima bean	44.6
	broad beans Vicia faba	pods, seeds	North Africa South and southwest Asia	broad bean	

	peas Pisum sativum	pods, seeds, sprouting	Mediterranean and Middle East	pea, snap pea, snow pea, split pea	28.9
	potato Solanum tuberosum	root tubers	South America	potato	365.4
	aubergine/egg-plant Solanum melongena	fruits	South and East Asia	eggplant (aubergine)	48.4
	tomato Solanum lycopersicum	fruits	South America	tomato	161.8
	cucumber Cucumis sativus	fruits	Southern Asia	cucumber	65.1
	pumpkin/squash Cucurbita spp.	fruits, flowers	Mesoamerica	pumpkin, squash, marrow, zucchini (courgette), gourd	24.6
	onion Allium cepa	bulbs, leaves	Asia	onion, spring onion, scallion, shallot	87.2
	garlic Allium sativum	bulbs	Asia	garlic	24.8
	leek Allium ampeloprasum	leaf sheaths	Europe and Middle East	leek, elephant garlic	21.7

	pepper Capsicum annuum	fruits	North and South America	pepper, bell pepper, sweet pepper	34.5
	spinach Spinacia oleracea	leaves	Central and southwestern Asia	spinach	21.7
	yam Dioscorea spp.	tubers	Tropical Africa	yam	59.5
	sweet potato Ipomoea batatas	tubers, leaves, shoots	Central and South America	sweet potato	108.0
	cassava Manihot esculenta	tubers	South America	cassava	269.1

1. Includes both carrots and turnips.

2. Productions of dry and green vegetables added up.

Nutrition and Health

Southeast Asian style stir fry ipomoea aquatica in chili and sambal.

Vegetables play an important role in human nutrition. Most are low in fat and calories but are bulky and filling. They supply dietary fibre and are important sources of essential vitamins, minerals, and trace elements. Particularly important are the antioxidant vitamins A, C, and E. When vegetables are included in the diet, there is found to be a reduction in the incidence of cancer, stroke, cardiovascular disease, and other chronic ailments. Research has shown that, compared with individuals who eat less than three servings of fruits and vegetables each day, those that eat more than five servings have an approximately twenty percent lower risk of developing coronary heart disease or stroke. The nutritional content of vegetables varies considerably; some contain useful amounts of protein though generally they contain little fat, and varying proportions of vitamins such as vitamin A, vitamin K, and vitamin B_6; provitamins; dietary minerals; and carbohydrates. Vegetables contain a great variety of other phytochemicals (bioactive non-nutrient plant compounds), some of which have been claimed to have antioxidant, antibacterial, antifungal, antiviral, and anticarcinogenic properties.

Vegetables (and some fruit) for sale on a street in Guntur, India.

However, vegetables often also contain toxins and antinutrients which interfere with the absorption of nutrients. These include α-solanine, α-chaconine, enzyme inhibitors (of cholinesterase, protease, amylase, etc.), cyanide and cyanide precursors, oxalic acid, and others. These toxins are natural defenses, used to ward off the insects, predators and fungi that might attack the plant. Some beans contain phytohaemagglutinin, and cassava roots contain cyanogenic glycoside as do bamboo shoots. These toxins can be deactivated by adequate cooking. Green potatoes contain glycoalkaloids and should be avoided.

Fruit and vegetables, particularly leafy vegetables, have been implicated in nearly half the gastrointestinal infections caused by norovirus in the United States. These foods are commonly eaten raw and may become contaminated during their preparation by an infected food handler. Hygiene is important when handling foods to be eaten raw, and such products need to be properly cleaned, handled, and stored to limit contamination.

Dietary Recommendations

The USDA Dietary Guidelines for Americans recommends consuming five to nine servings of fruit and vegetables daily. The total amount consumed will vary according to age and gender, and is determined based upon the standard portion sizes typically consumed, as well as general nutritional content. Potatoes are not included in the count as they are mainly providers of starch. For most vegetables and vegetable juices, one serving is half of a cup and can be eaten raw or cooked. For leafy greens, such as lettuce and spinach, a single serving is typically a full cup. A variety of products should be chosen as no single fruit or vegetable provides all the nutrients needed for health.

International dietary guidelines are similar to the ones established by the USDA. Japan, for example, recommends the consumption of five to six servings of vegetables daily. French recommendations provide similar guidelines and set the daily goal at five servings. In India, the daily recommendation for adults is 275 grams (9.7 oz) of vegetables per day.

Production

Cultivation

Growing vegetables in South Africa.

Vegetables have been part of the human diet from time immemorial. Some are staple foods but most are accessory foodstuffs, adding variety to meals with their unique flavors and at the same time, adding nutrients necessary for health. Some vegetables are perennials but most are annuals and biennials, usually harvested within a year of sowing or planting. Whatever system is used for growing crops, cultivation follows a similar pattern; preparation of the soil by loosening it, removing or burying weeds, and adding organic manures or fertilisers; sowing seeds or planting young plants; tending the crop while it grows to reduce weed competition, control pests, and provide sufficient water; harvesting the crop when it is ready; sorting, storing, and marketing the crop or eating it fresh from the ground.

Different soil types suit different crops, but in general in temperate climates, sandy soils dry out fast but warm up quickly in the spring and are suitable for early crops, while heavy clays retain moisture better and are more suitable for late season crops. The growing season can be lengthened by the use of fleece, cloches, plastic mulch, polytunnels, and greenhouses. In hotter regions, the production of vegetables is constrained by the climate, especially the pattern of rainfall, while in temperate zones, it is constrained by the temperature and day length.

Weeding cabbages in Colorado, US.

On a domestic scale, the spade, fork, and hoe are the tools of choice while on commercial farms a range of mechanical equipment is available. Besides tractors, these include ploughs, harrows, drills, transplanters, cultivators, irrigation equipment, and harvesters. New techniques are changing the cultivation procedures involved in growing vegetables with computer monitoring systems, GPS locators, and self-steer programs for driverless machines giving economic benefits.

Harvesting

Harvesting beetroot in the United Kingdom.

When a vegetable is harvested, it is cut off from its source of water and nourishment. It continues to transpire and loses moisture as it does so, a process most noticeable in the wilting of green leafy crops. Harvesting root vegetables when they are fully mature improves their storage life, but alternatively, these root crops can be left in the ground and harvested over an extended period. The harvesting process should seek to minimise damage and bruising to the crop. Onions and garlic can be dried for a few days in the field and root crops such as potatoes benefit from a short maturation period in warm, moist surroundings, during which time wounds heal and the skin thickens up and hardens. Before marketing or storage, grading needs to be done to remove damaged goods and select produce according to its quality, size, ripeness, and color.

Storage

All vegetables benefit from proper post harvest care. A large proportion of vegetables and perishable foods are lost after harvest during the storage period. These losses may be as high as thirty to fifty percent in developing countries where adequate cold storage facilities are not available. The main causes of loss include spoilage caused by moisture, moulds, micro-organisms, and vermin.

Temporary storage of potatoes in the Netherlands.

Storage can be short-term or long-term. Most vegetables are perishable and short-term storage for a few days provides flexibility in marketing. During storage, leafy vegetables lose moisture, and the vitamin C in them degrades rapidly. A few products such as potatoes and onions have better keeping qualities and can be sold when higher prices may be available, and by extending the marketing season, a greater total volume of crop can be sold. If refrigerated storage is not available, the priority for most crops is to store high-quality produce, to maintain a high humidity level, and to keep the produce in the shade.

Proper post-harvest storage aimed at extending and ensuring shelf life is best effected by efficient cold chain application. Cold storage is particularly useful for vegetables such as cauliflower, eggplant, lettuce, radish, spinach, potatoes, and tomatoes, the optimum temperature depending on the type of produce. There are temperature-controlling technologies that do not require the use of electricity such as evaporative cooling. Storage of fruit and vegetables in controlled atmospheres with high levels of carbon dioxide or high oxygen levels can inhibit microbial growth and extend storage life.

The irradiation of vegetables and other agricultural produce by ionizing radiation can be used to preserve it from both microbial infection and insect damage, as well as from physical deterioration. It can extend the storage life of food without noticeably changing its properties.

Preservation

The objective of preserving vegetables is to extend their availability for consumption or marketing purposes. The aim is to harvest the food at its maximum state of palatability and nutritional value, and preserve these qualities for an extended period. The main causes of deterioration in vegetables after they are gathered are the actions of naturally-occurring enzymes and the spoilage caused by micro-organisms. Canning and freezing are the most commonly used techniques, and vegetables preserved by these methods are generally similar in nutritional value to comparable fresh products with regards to carotenoids, vitamin E, minerals. and dietary fiber.

Bean field and canning factory, New Jersey, US.

Canning is a process during which the enzymes in vegetables are deactivated and the micro-organisms present killed by heat. The sealed can excludes air from the foodstuff to prevent subsequent deterioration. The lowest necessary heat and the minimum processing time are used in order to prevent the mechanical breakdown of the product and to preserve the flavor as far as is possible. The can is then able to be stored at ambient temperatures for a long period.

Freezing vegetables and maintaining their temperature at below −10 °C (14 °F) will prevent their spoilage for a short period, whereas a temperature of −18 °C (0 °F) is required for longer-term storage. The enzyme action will merely be inhibited, and blanching of suitably sized prepared vegetables before freezing mitigates this and prevents off-flavors developing. Not all micro-organisms will be killed at these temperatures and after thawing the vegetables should be used promptly because otherwise, any microbes present may proliferate.

Sun-drying tomatoes in Greece.

Traditionally, sun drying has been used for some products such as tomatoes, mushrooms, and beans, spreading the produce on racks and turning the crop at intervals. This method suffers from several disadvantages including lack of control over drying rates, spoilage when drying is slow, contamination by dirt, wetting by rain, and attack by rodents, birds, and insects. These disadvantages can be alleviated by using solar powered driers. The dried produce must be prevented from reabsorbing moisture during storage.

High levels of both sugar and salt can preserve food by preventing micro-organisms from growing. Green beans can be salted by layering the pods with salt, but this method of preservation is unsuited to most vegetables. Marrows, beetroot, carrot, and some other vegetables can be boiled with sugar to create jams. Vinegar is widely used in food preservation; a sufficient concentration of acetic acid prevents the development of destructive micro-organisms, a fact made use of in the preparation of pickles, chutneys and relishes. Fermentation is another method of preserving vegetables for later use. Sauerkraut is made from chopped cabbage and relies on lactic acid bacteria which produce compounds that are inhibitory to the growth of other micro-organisms.

Top Producers

Farmers' market showing vegetables for sale near the Potala Palace in Lhasa, Tibet.

Vegetable shop in India.

Vegetables in a supermarket in the United States.

Vegetables in a supermarket in Canada.

In 2010, China was the largest vegetable producing nation, with over half the world's production. India, the United States, Turkey, Iran, and Egypt were the next largest producers. China had the highest area of land devoted to vegetable production, while the highest average yields were obtained in Spain and the Republic of Korea.

Country	Area cultivated thousand hectares (2,500 acres)	Yield thousand kg/ha (89 lbs/acre)	Production thousand tonnes (1,100 short tons)
China	23,458	230	539,993
India	7,256	138	100,045
United States	1,120	318	35,609
Turkey	1,090	238	25,901
Iran	767	261	19,995
Egypt	755	251	19,487
Italy	537	265	14,201
Russia	759	175	13,283
Spain	348	364	12,679
Mexico	681	184	12,515
Nigeria	1844	64	11,830
Brazil	500	225	11,233
Japan	407	264	10,746
Indonesia	1082	90	9,780
South Korea	268	364	9,757
Vietnam	818	110	8,976
Ukraine	551	162	8,911
Uzbekistan	220	342	7,529
Philippines	718	88	6,299
France	245	227	5,572
Total world	55,598	188	1,044,380

Standards

The International Organization for Standardization (ISO) sets international standards to ensure that products and services are safe, reliable, and of good quality. There are a number of ISO standards regarding fruits and vegetables. ISO 1991-1:1982 lists the botanical names of sixty-one species of plants used as vegetables along with the common names of the vegetables in English, French, and Russian. ISO 67.080.20 covers the storage and transport of vegetables and their derived products.

Murcia

Murcia is a city in south-eastern Spain, the capital and most populous city of the Autonomous Community of the Region of Murcia, and the seventh largest city in the country, with a population of 442,573 inhabitants in 2009 (about one third of the total population of the Region). The population of the metropolitan area was 689,591 in 2010. It is located on the Segura River, in the Southeast of the Iberian Peninsula, noted by a climate with hot summers, mild winters, and relatively low precipitation.

Murcia was founded by the emir of Cordoba Abd ar-Rahman II in 825 with the name Mursiyah مرسية and nowadays is mainly a services city and a university town. Highlights for visitors include the Cathedral of Murcia and a number of baroque buildings, renowned local cuisine, Holy Week procession works of art by the famous Murcian sculptor Francisco Salzillo, and the Fiestas de Primavera (Spring Festival).

The city, as the capital of the comarca Huerta de Murcia is called Europe's orchard due to its long agricultural tradition and its fruit, vegetable, and flower production and exports.

Geography

Murcia is located near the center of a low-lying fertile plain known as the huerta (orchard or vineyard) of Murcia. The Segura River and its right-hand tributary, the Guadalentín, run through the area. The city has an elevation of 43 metres (141 ft) above sea level and covers approximately 882 square kilometres (341 sq mi).

The best known and most dominant aspect of the municipal area's landscape is the orchard. In addition to the orchard and urban zones (Alfonso X, Gran Via, Jaime I, and others), the great expanse of the municipal area is made up of different landscapes: badlands, groves of Carrasco pine trees in the precoastal mountain ranges and, towards the south, a semi-steppe region.

A large regional park, the Parque Regional de Carrascoy y el Valle, lies just to the south of the city.

Segura River

Murcia is located in the Segura valley.

The Segura River crosses an alluvial plain (Vega Media del Segura), part of a Mediterranean pluvial system. The river crosses the city from west to east. Its volumetric flow is mostly small but the river is known to produce occasional flooding, like those that inundated the capital in 1946, 1948, 1973 or 1989. The Segura was recognized as one of the most polluted rivers in Europe.

Mountains and Hills

The Segura river's Valley is surrounded by two mountain ranges, the hills of Guadalupe, Espinardo, Cabezo de Torres, Esparragal and Monteagudo in the north and the Cordillera Sur in the south. The municipality itself is divided into southern and northern zones by a series of mountain ranges, the aforementioned Cordillera Sur (Carrascoy, El Puerto, Villares, Columbares, Altaona, and Escalona). These two zones are known as Field of 'Murcia (in the south of Cordillera Sur) and Orchard of Murcia (the Segura Valley in the north of Cordillera Sur). Near the plain's center, the steep hill of Monteagudo protrudes dramatically.

Districts

The 881.86-square-kilometre (340.49 sq mi) territory of Murcia's municipality is made up of 54 pedanías (suburban districts) and 28 barrios (city neighbourhood districts). The barrios make up the 12.86-square-kilometre (4.97 sq mi) the main urban portion of the city. The historic city center is approximately 3 square kilometres (1 sq mi) of the urbanized downtown portion of Murcia.

Climate

Murcia has a hot semi-arid climate (Köppen climate classification: BSh), with arid climate (BWh) influences. Given its proximity to the Mediterranean Sea, it has mild winters and hot summers.

It averages more than 320 days of sun per year. Occasionally, Murcia has heavy rains where the precipitation for the entire year will fall over the course of a few days.

In the coldest month, January, the average temperature range is a high of 16.6 °C (62 °F) during the day and a low of 4.7 °C (40 °F) at night. In the warmest month, August, the range goes from 34.2 °C (94 °F) during the day to 20.9 °C (70 °F) at night. Temperatures almost always reach or exceed 40 °C (104 °F) on at least one or two days per year. In fact, Murcia holds temperature records close to the highest recorded in southern Europe since reliable meteorological records commenced in 1950. The official record for Murcia stands at a stifling 47.2 °C (117.0 °F), at Alcantarilla airport in the western suburbs on July 4, 1994 with 45.7 °C (114.3 °F) being recorded at a station near the city centre on the same day.

Climate data for Murcia (1981–2010)													
Month	Jan	Feb	Mar	Apr	May	Jun	Jul	Aug	Sep	Oct	Nov	Dec	Year
Record high °C (°F)	25.8 (78.4)	29.4 (84.9)	32.6 (90.7)	37.4 (99.3)	38.5 (101.3)	42.5 (108.5)	47.2 (117)	43.2 (109.8)	44.6 (112.3)	34.9 (94.8)	31.0 (87.8)	25.8 (78.4)	47.2 (117)
Average high °C (°F)	16.6 (61.9)	18.4 (65.1)	20.9 (69.6)	23.3 (73.9)	26.6 (79.9)	31.0 (87.8)	34.0 (93.2)	34.2 (93.6)	30.4 (86.7)	25.6 (78.1)	20.3 (68.5)	17.2 (63)	24.9 (76.8)
Daily mean °C (°F)	10.6 (51.1)	12.2 (54)	14.3 (57.7)	16.5 (61.7)	20.0 (68)	24.2 (75.6)	27.2 (81)	27.6 (81.7)	24.2 (75.6)	19.8 (67.6)	14.6 (58.3)	11.5 (52.7)	18.6 (65.5)
Average low °C (°F)	4.7 (40.5)	5.9 (42.6)	7.7 (45.9)	9.7 (49.5)	13.3 (55.9)	17.4 (63.3)	20.3 (68.5)	20.9 (69.6)	18.0 (64.4)	13.9 (57)	8.9 (48)	5.8 (42.4)	12.3 (54.1)
Record low °C (°F)	−7.5 (18.5)	−3.9 (25)	−2.4 (27.7)	0.0 (32)	4.0 (39.2)	8.0 (46.4)	13.0 (55.4)	14.0 (57.2)	9.6 (49.3)	4.4 (39.9)	−1.0 (30.2)	−6.0 (21.2)	−7.5 (18.5)
Average precipitation mm (inches)	27 (1.06)	27 (1.06)	30 (1.18)	25 (0.98)	28 (1.1)	18 (0.71)	3 (0.12)	8 (0.31)	32 (1.26)	36 (1.42)	32 (1.26)	29 (1.14)	297 (11.69)
Average precipitation days (≥ 1 mm)	4	4	3	4	4	2	1	1	3	4	4	4	37
Mean monthly sunshine hours	189	190	223	256	289	323	353	317	239	217	186	172	2,967

Climate data for Murcia—San Javier (Airport 4 m, near sea) (1981–2010)												
Month	Jan	Feb	Mar	Apr	May	Jun	Jul	Aug	Sep	Oct	Nov	Dec	Year
Record high °C (°F)	26.2 (79.2)	27.8 (82)	30.0 (86)	32.0 (89.6)	34.5 (94.1)	36.9 (98.4)	40.5 (104.9)	40.0 (104)	39.4 (102.9)	35.5 (95.9)	30.0 (86)	27.0 (80.6)	40.5 (104.9)
Average high °C (°F)	16.0 (60.8)	16.7 (62.1)	18.5 (65.3)	20.4 (68.7)	22.9 (73.2)	26.4 (79.5)	28.9 (84)	29.5 (85.1)	27.5 (81.5)	24.0 (75.2)	19.8 (67.6)	17.6 (63.7)	22.3 (72.1)
Daily mean °C (°F)	10.8 (51.4)	11.6 (52.9)	13.4 (56.1)	15.3 (59.5)	18.4 (65.1)	22.2 (72)	24.8 (76.6)	25.5 (77.9)	23.2 (73.8)	19.4 (66.9)	14.9 (58.8)	11.9 (53.4)	17.6 (63.7)
Average low °C (°F)	5.5 (41.9)	6.5 (43.7)	8.4 (47.1)	10.2 (50.4)	13.8 (56.8)	17.9 (64.2)	20.7 (69.3)	21.5 (70.7)	18.9 (66)	14.7 (58.5)	10.0 (50)	6.8 (44.2)	12.9 (55.2)
Record low °C (°F)	−3.8 (25.2)	−4.0 (24.8)	−3.0 (26.6)	1.0 (33.8)	4.8 (40.6)	9.5 (49.1)	11.0 (51.8)	12.0 (53.6)	7.9 (46.2)	4.0 (39.2)	−1.5 (29.3)	−5.4 (22.3)	−5.4 (22.3)
Average precipitation mm (inches)	42 (1.65)	27 (1.06)	24 (0.94)	23 (0.91)	25 (0.98)	7 (0.28)	2 (0.08)	7 (0.28)	39 (1.54)	39 (1.54)	47 (1.85)	30 (1.18)	313 (12.32)
Average precipitation days (≥ 1 mm)	4	3	3	3	3	1	0	1	3	4	4	4	33
Mean monthly sunshine hours	173	171	206	224	266	288	307	283	224	200	162	156	2,621

History

Muslim architecture of the Alcázar Seguir in Santa Clara Museum inside of Monasterio de Santa Clara la Real.

It is widely believed that Murcia's name is derived from the Latin words of Myrtea or Murtea, meaning land of Myrtle (the plant is known to grow in the general area), although it may also be a derivation of the word Murtia, which would mean Murtius Village (Murtius was a common Roman name). Other research suggests that it may owe its name to the Latin Murtae (Mulberry), which covered the regional landscape for many centuries. The Latin name eventually changed into the Arabic Mursiya, and then, Murcia.

Entrance of James I of Aragon at Murcia in 1266.

The city in its present location was founded with the name Madinat Mursiyah (city of Murcia) in AD 825 by Abd ar-Rahman II, who was then the emir of Córdoba. Muslim planners, taking advantage of the course of the river Segura, created a complex network of irrigation channels that made the town's agricultural existence prosperous. In the 12th century the traveler and writer Muhammad al-Idrisi described the city of Murcia as populous and strongly fortified. After the fall of the Caliphate of Cordoba in 1031, Murcia passed under the successive rules of the powers seated variously at Almería, Toledo and Seville. After the fall of Almoravide empire, Muhammad Ibn Mardanis made

Murcia the capital of an independent kingdom. At this time, Murcia was a very prosperous city, famous for its ceramics, exported to Italian towns, as well as for silk and paper industries, the first in Europe. The coinage of Murcia was considered as model in all the continent. The mystic Ibn Arabi (1165–1240) and the poet Ibn al-Jinan (d.1214) were born in Murcia during this period.

In 1172 Murcia was conquered by the north African based Almohades, the last Muslim empire to rule southern Spain, and as the forces of the Christian Reconquista gained the upper hand, was the capital of a small Muslim emirate from 1223 to 1243. By the treaty of Alcaraz, in 1243, the Christian king Ferdinand III of Castile made Murcia a protectorate, getting access to the Mediterranean sea while Murcia was protected against Granada and Aragon. The Christian population of the town became the majority as immigrants poured in from almost all parts of the Iberian Peninsula. Christian immigration was encouraged with the goal of establishing a loyal Christian base. These measures led to the Muslim population revolt in 1264, which was quelled by James I of Aragon in 1266, bringing Aragonese and Catalonian immigrants with him.

After this, during the reign of Alfonso X of Castile, Murcia was one of his capitals with Toledo and Seville.

The Murcian duality: Catalonian population in a Castillian territory, brought the subsequent conquest of the city by James II of Aragon in 1296. In 1304, Murcia was finally incorporated into Castile under the Treaty of Torrellas.

Murcia Flood in 1879.

Murcia's prosperity declined as the Mediterranean lost trade to the ocean routes and from the wars between the Christians and the Ottoman Empire. The old prosperity of Murcia became crises during 14th century because of its border location with the neighbouring Muslim kingdom of Granada, but flourished after its conquest in 1492 and again in the 18th century, benefiting greatly from a boom in the silk industry. Most of the modern city's landmark churches, monuments and old architecture date from this period. In this century, Murcia lived an important role in Bourbon victory in the War of the Spanish Succession, thanks to the Cardinal Belluga. In 1810, Murcia was looted by Napoleonic troops; it then suffered a major earthquake in 1829. According to contemporaneous accounts, an estimated 6,000 people died from the disaster's effects across the province. Plague and cholera followed.

The town and surrounding area suffered badly from floods in 1651, 1879, and 1907, though the construction of a levee helped to stave off the repeated floods from the Segura. A popular pedestrian walkway, the Malecon, runs along the top of the levee.

Murcia has been the capital of the province of Murcia since 1833 and, with its creation by the central government in 1982, capital of the autonomous community (which includes only the city and the province). Since then, it has become the seventh most populated municipality in Spain, and a thriving services city.

On May 11, 2011, the city of Lorca and surrounding area was struck by a 5.3 magnitude earthquake. At least 4 people were reported to have died as a result of the earthquake.

Demographics

Murcia Cathedral of Santa Maria.

Al-Andalusian palatial complex and neighborhood of San Esteban.

The town hall.

Murcia has 433,850 inhabitants (INE 2008) making it the seventh-largest Spanish municipality by population. When adding in the municipalities of Alcantarilla, Alguazas, Beniel, Molina de Segura, Santomera, and Las Torres de Cotillas, the metropolitan area has 564,036 inhabitants making it the twelfth most populous metropolitan area in

Spain. Nevertheless, due to Murcia's large municipal territory, its population density (472 hab./km², 760 hab./sq.mi.) does not likewise rank among Spain's highest.

According to the official population data of the INE, 10% of the population of the municipality reported belonging to a foreign nationality as of 2005.

The majority of the population identify as Christian. There is also a sizeable Muslim population as well as a growing Jewish community.

Main Sights

The Cathedral of Murcia was built between 1394 and 1465 in the Castilian Gothic style. Its tower was completed in 1792 and shows a blend of architectural styles. The first two stories were built in the Renaissance style (1521–1546), while the third is Baroque. The bell pavilion exhibits both Rococo and Neoclassical influences. The main façade (1736–1754) is considered a masterpiece of the Spanish Baroque style.

Other noteworthy buildings in the square shared by the Cathedral (Plaza Cardinal Belluga) are the colorful Bishop's Palace (18th century) and a controversial extension to the town hall by Rafael Moneo (built in 1999).

The Glorieta, which lies on the banks of the Segura River, has traditionally been the center of the town. It is a pleasant, landscaped city square that was constructed during the 18th century. The ayuntamiento (city hall) of Murcia is located in this square.

Pedestrian areas cover most of the old town of the city, which is centered around Platería and Trapería Streets. Trapería goes from the Cathedral to the Plaza de Santo Domingo, formerly a bustling market square. Located in Trapería is the Casino, a social club erected in 1847, with a sumptuous interior that includes a Moorish-style patio inspired by the royal chambers of the Alhambra near Granada. The name Plateria refers to plata (silver), as this street was the historical focus for the commerce of rare metals by Murcia's Jewish community. The other street, Traperia, refers to trapos, or cloths, as this was once the focus for the Jewish community's garment trade.

Several bridges of different styles span the river Segura, from the Puente de los Peligros, eighteenth century stone bridge with a Lady chapel on one of its sides; to modern bridges designed by Santiago Calatrava or Javier Manterola; through others such as the Puente Nuevo, an iron bridge of the early twentieth century.

Other notable places around Murcia include:

- Santa Clara monastery, a Gothic and Baroque monument where is located a museum with the Moorish palace's remains from the 13th century, called Alcázar Seguir.

- The Malecón boulevard, a former retaining wall for the Río Segura's floods.

- La Fuensanta sanctuary and adjacent El Valle regional park.

- Los Jerónimos monastery (18th century).

- Romea theatre (19th century).

- Almudí Palace (17th century), a historic building with coats of arms on its façade. On its interior there are Tuscan columns, and since 1985 it hosts the city archives and usually houses exhibitions.

- Monteagudo Castle (11th century).

- Salzillo Museum.

- San Juan de Dios church-museum, Baroque and Rococo circular church with the remains of the Moorish palace mosque from the 12th century in the basament, called Alcázar Nasir.

In the metropolitan area are also the Azud de la Contraparada reservoir and the Noria de La Ñora water wheel.

Festivals

The Burial of the Sardine in Murcia.

The Holy Week procession hosted by the city is among the most famous throughout Spain. This traditional festival portrays the events which lead up to and include the Crucifixion according to the New Testament. Life-sized, finely detailed sculptures by Francisco Salzillo (1707–1783) are removed from their museums and carried around the city in elegant processions amid flowers and, at night, candles, pausing at stations which are meant to re-enact the final moments before the crucifixion of Jesus.

The most colorful festival in Murcia may come one week after Holy Week, when locals dress up in traditional huertano clothing to celebrate the Bando de la Huerta (Orchard parade) on Tuesday and fill the streets for The Burial of the Sardine in Murcia. parade the following Saturday. This whole week receives the name of Fiestas de Primavera (Spring Fest).

Murcia's Three Cultures International Festival happens each May and was first organized with the intent of overcoming racism and xenophobia in the culture. The festival seeks to foster understanding and reconciliation between the three cultures that have cohabited the peninsula for centuries, if not millennia: Christians, Jews and Muslims. Each year, the festival celebrates these three cultures through music, exhibitions, symposiums and conferences.

Economy

Trapería Street in the old town.

Economically, Murcia predominantly acts as a centre for agriculture and tourism. It is common to find Murcia's tomatoes and lettuce, and especially lemons and oranges, in European supermarkets. Murcia is a producer of wines, with about 40,000 hectares (100,000 acres) devoted to grape vineyards. Most of the vineyards are located in Ricote and Jumilla. Jumilla is a plateau where the vineyards are surrounded by mountains.

Murcia has some industry, with foreign companies choosing it as a location for factories, such as Henry Milward & Sons (which manufactures surgical and knitting needles) and American firms like General Electric and Paramount Park Studios.

During the 2000s, the economy of the region turned towards "residential tourism" in which people from northern European countries have a second home in the area. Europeans and Americans are able to learn Spanish in the academies in the town center.

The economy of Murcia is supported by fairs and congresses, museums, theatres, cinema, music, aquariums, bullfighting, restaurants, hotels, camping, sports, foreign students, and tourism.

Transportation

Tram of Murcia.

By Plane

Murcia-San Javier Airport (MJV) is located on the edge of the Mar Menor close to the town of San Javier, 45 kilometres (28 miles) southeast of Murcia. There is also an airport at the neighboring city of Alicante 70 km (43 miles) from Murcia. Furthermore, there is a new airport in development to be located in the town of Corvera, 23 km (14 miles).

By Bus

Bus service is provided by LatBus, which operates the interurban services. Urban bus services is offered by a new operator, TM(Transportes de Murcia), an UTE (Joint Venture) formed by Ruiz, Marín & Fernanbús.

By Tram

Tramways are managed by Tranvimur. As of 2007, 2 kilometres (1 mile) of line were available. Since 2011, one line is connecting the city center (Plaza Circular) with the University Campus and the Football Stadium.

By Train

Train connections are provided by RENFE. Murcia has a railway station called Murcia del Carmen, located in the neighborhood of the same name. Several long-distance lines link the city with Madrid, through Albacete, as well as Valencia, and Cataluña up to Montpellier in France. Murcia is also the center of a local network. The line C-1 connects the city to Alicante, and the line C-2 connects Murcia to Alcantarilla, Lorca and Águilas. It also has two regional lines connecting it to Cartagena and Valencia.

Healthcare

The hospitals and other public primary healthcare centers belong to the Murcian Healthcare Service. There are three public hospitals in Murcia:

- Ciudad Sanitaria Virgen de La Arrixaca in El Palmar that includes obstetrics and paediatrics units.

- Hospital Reina Sofía.

- Hospital Morales Meseguer.

Education

University of Murcia.

University of Murcia (cloister).

Murcia has two universities:

- One public university: the University of Murcia, founded in 1272.

- One private university: the Catholic University Saint Anthony, founded in 1996.

There are several high schools, elementary schools, and professional schools. Murcia has three types of schools for children: private schools such as El Limonar International School, Murcia (an American international school) and King's College, Murcia (a British international school), semi-private schools (concertado), which are private schools that receive government funding and sometimes offer religious instruction, and public schools such as Colegio Publico (CP) San Pablo or the centenary CP Cierva Peñafiel, one of the oldest ones. The French international school, Lycée Français André Malraux de Murcie, is in nearby Molina de Segura.

The private schools and concertados can be religious (Catholic mostly but any religion is acceptable) or secular, but the public schools are strictly secular. Concertado or semi-private or quasi-private schools fill a need by providing schools where the government isn't able to or predate the national school system.

Instituto Licenciado Cascales is one of the oldest in the city, built in 1724, and perhaps the most emblematic. IES Alfonso X El Sabio is the only school in Murcia to offer the prestigious International High School Diploma.

Murcia also offers Adult Education for people who want to return to complete high school and possibly continue on to the university.

Sports Teams

- Real Murcia: Spanish Third Division football.
- CF Atlético Ciudad: Spanish Third Division (Group 2) football—dissolved in 2010.
- CB Murcia: Liga ACB basketball.
- ElPozo Murcia Turística FS: futsal.
- The Hispania Racing F1 Team is also based in Murcia, and receives sponsorship from the tourist board.
- CAV Murcia 2005: Superliga Femenina de Voleibol volleyball.
- Origen (esports): League of Legends team.

Twin Towns—Sister Cities

Murcia is twinned with:

- Lecce, Italy
- Grasse, France
- Irapuato, Mexico
- Murcia, Philippines
- Łódź, Poland, since 1999

Tenbury Wells

Tenbury Wells (locally Tenbury) is a market town and civil parish in the north-western extremity of the Malvern Hills District of Worcestershire, England, which at the 2011 census had a population of 3,777.

Geography

Tenbury Wells lies on the south bank of the River Teme, which forms the border between Shropshire and Worcestershire. It is in the north-west of the Malvern Hills District. The settlement of Burford in Shropshire lies on the north bank of the river.

History

From 1894 to 1974, it was a rural district, comprising itself and villages such as Stoke Bliss, Eastham and Rochford. From 1974 Tenbury was in the District of Leominster until it became part Malvern Hills District when Leominster District Council was taken over by Herefordshire Council in April 1998.

The history of Tenbury Wells extends as far back as the Iron Age. The town is often thought of as the home to the Castle Tump, but this is now in Burford, Shropshire due to boundary changes. Though the Tump, possibly the remains of an early Norman motte and bailey castle, can be seen from the main road (A456) there are no visible remains of the castle that was constructed to defend and control the original River Teme crossing. It has also been described as "... the remains of an 11th century Norman Castle."

A legal record of 1399 mentions a place spelt perhaps as Temedebury which may be a further variation in spelling.

Tenbury was in the upper division of Doddingtree Hundred.

Originally named "Temettebury", the town was granted a Royal Charter to hold a market in 1249. Over time, the name changed to "Tenbury", and then added the "Wells" following the discovery of mineral springs and wells in the town in the 1840s. The name of the railway station, which was on the now-defunct Tenbury & Bewdley Railway, was changed in 1912, in an attempt to publicise the mineral water being produced from the wells around the town.

The St Michael and All Angels Choir School devoted to the Anglican choral tradition by Frederick Ouseley closed in 1985 and the buildings now serve alternative educational purposes.

For over 100 years Tenbury has been well known throughout the country for its winter auctions of holly and mistletoe (and other Christmas products). It is also known for its "Chinese-gothic" Pump Room buildings, built in 1862, which re-opened in 2001, following a major restoration. They are now owned by Tenbury Town Council, having been transferred from Malvern Hills District Council in September 2008.

Architecture

Eastham Bridge near Tenbury, which collapsed in May 2016.

One notable architectural feature in the town is the unique (often described as Chinese-Gothic) Pump Rooms, designed by James Cranston in the 1860s, to house baths where the mineral water was available.

Other notable structures in Tenbury include the parish church of St Mary with a Norman tower, and a number of monuments. The church was essential rebuilt by Henry Woodyer between 1864 and 1865.

The part-medieval bridge over the River Teme, linking Tenbury to Burford, Shropshire was rebuilt by Thomas Telford following flood damage in 1795.

The Grade II-listed Eastham bridge dramatically collapsed into the River Teme on 24 May 2016. There were no reports of any casualties.

The Victorian Workhouse, designed by George Wilkinson, was used as the local Council Buildings from 1937 to the early 21st century and is currently being converted into residential housing. The Victorian infirmary behind the workhouse is scheduled to be demolished to create car parking for a new Tesco Superstore.

The unique Victorian corrugated iron isolation hospital was demolished on 24 October 2006.

Local Interest

Markets

Markets are held on Tuesday mornings, Friday mornings, and Saturday mornings, in and around the town's Round Market building, which was built by James Cranston in 1858. In 2013, a new monthly 'local producers market' started, initially held near the Pump Rooms, more recently on Teme Street.

Apple and Fruit Heritage

Tenbury was also known as "the town in the orchard" due to the large numbers of fruit orchards of apple trees and also pears, quince and plum trees, in the immediate vicinity of the town. This heritage is revisited every October during the Tenbury Applefest. Tenbury Applefest website.

Tenbury in Poetry

Orchards gay with blossom,
Beauty, there to see,
Hollows where breeze is tender,
Moorlands where wind breaks free;
Sowing, Lambing, and Harvest,
Overlooked by Giant Clee,
Hop Kilns, Farmsteads, and TENBURY,
This is happiness for me;

Power Station Shelved

A proposal to build a biomass power station on a business park failed due to residents' concern about the disruption to local businesses during its construction. The proposal continued to attract protests, and in July 2007 a petition against the plans was signed by more than 2,300 people. In July 2009 it was announced that the £965,000 grant offered to the power station had been withdrawn and the project shelved.

Local Flooding

For several centuries Tenbury has been subject to regular flooding on many occasions, and most recently in 2007 and in 2008.The first flood was caused by the River Teme and the Kyre Brook bursting their banks. The second was caused by a combination of 15mm (0.59 in) of rain falling in an hour and the town's drainage system (much of which was blocked) failing to cope, creating flash flooding. The third flood again involved the River Teme and the Kyre Brook bursting their banks. The 2008 flood damage was caused by a combination of the drainage not having been upgraded since the 2007 floods and the wall on Market Street (which should hold back the Kyre Brook) not having been rebuilt following the 2007 floods. Since then much work has been done in respect of improved drainage and particularly defences in Market Street.

Regal Cinema

The Regal Cinema on Teme Street in Tenbury Wells opened in 1937. It operated as a commercial cinema as one of six in the Craven Cinemas chain, until the decline of British cinema led to its closure in 1966. Following purchase by Tenbury Town Council to prevent demolition, various volunteer groups have run it.

The Regal has been subject of a Heritage Lottery Fund supported restoration project. Replicas of the 1930s mediterranean murals by artist George Legge have been painted around the auditorium, the detailing on the front of the building has been recreated, and neon lighting has been erected on the front canopy. The building, owned by Tenbury Town Council is now under the management of a trust. Modern equipment now allows the showing of recently released films, live broadcasts and live acts. Paul Daniels was its patron until his death.

In 2016 The Regal in Tenbury has been nominated for the "Britain Has Spirit" award. An award that could see The Regal win £1,000 to host a street party for the Tenbury community if they win the regional vote, and possibly £25,000 if they win the national vote. The Regal are currently in the public voting stage and members of the public have up until the 16th June 2016 to vote for them to win. The competition is being run by Together Mutual Insurance

Education

For primary education Tenbury Wells is served by Tenbury CofE Primary School on Bromyard Road. Tenbury High Ormiston Academy on Oldwood Road is the main secondary school for the area, while King's St Michael's College (also on Oldwood Road) is an independent international boarding school.

Nearest Railway Stations

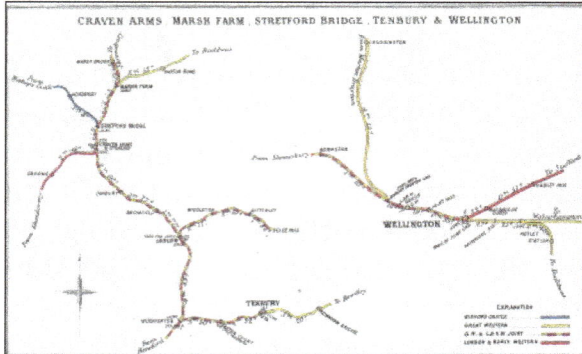

Railway Clearing House Junction Diagram of 1903. Woofferton railway station on the Welsh Marches Line.

The nearest open stations are located on the Welsh Marches Line are Ludlow railway station and Leominster.

The nearest point of operational railway is at Woofferton railway station, but it is currently closed.

Orchard House

Orchard House is a historic house museum in Concord, Massachusetts, USA. It was the longtime home of Amos Bronson Alcott (1799–1888) and his family, including his daughter Louisa May Alcott (1832-1888) who wrote and set her novel Little Women (1868–69) there.

History

The Alcotts had first moved to Concord in 1840, although they left in 1843 to start Fruitlands, a utopian agrarian commune in nearby Harvard. The family returned in 1845 and purchased a house named "Hillside", but left again in 1852, selling to Nathaniel Hawthorne who renamed it The Wayside.

The Alcotts returned to Concord once again in 1857. They moved into Orchard House, which was then two-story clapboard farmhouse, in the spring of 1858. At the time of purchase the site included two early eighteenth-century houses on a 12-acre (49,000m^2) apple orchard. Consequently, the Alcotts named it Orchard House. "'Tis a pretty retreat", Bronson Alcott wrote soon after moving in, "and ours; a family mansion to take pride in, rescued as it is from deformity and disgrace".

A. Bronson Alcott made significant changes to the building. He installed alcoves for busts retrieved from his failed Temple School, repaired the staircase, installed bookcases, constructed a back studio for his youngest daughter May's artwork, and installed a rustic fence around the property. He also moved a smaller tenant house to adjoin the rear of the main house, making a single larger structure. While the home was being renovated, the family rented rooms next door at The Wayside while the Hawthornes were living in England. Later, Lydia Maria Child visited the house and recorded her thoughts: "The result is a house full of queer nooks and corners and all manner of juttings in and out. It seems as if the spirit of some old architect had brought it from the Middle Ages and dropped it down in Concord... The whole house leaves a general impression of harmony, of a medieval sort".

Orchard House is adjacent to The Wayside on the historic "American Mile" roadway toward Lexington, and is less than a half-mile from Bush the home of Ralph Waldo Emerson, where Henry David Thoreau and the Alcotts were frequent visitors.

The Alcotts in Residence

Orchard House was the most permanent home of the Alcotts, with the family in residence from 1858 to 1877. During this period, the family included Bronson, his wife Abigail May, and their daughters Anna, Louisa, and May. Elizabeth, the model for Beth March, had died in March 1858, just weeks before the family moved in.

Orchard House, 1941.

The Alcott girls befriended the Hawthorne children, who lived next door, though Nathaniel Hawthorne himself was elusive. Bronson Alcott was disappointed and recorded, "Nobody gets a chance to speak with him unless by accident." However, he added, "Still he has a tender kindly side, and a voice that a woman might own, the hesitance is so taking, and the tones so remote from what you expected."

The Alcotts were vegetarians and harvested fruits and vegetables from the gardens and orchard on the property. Conversations about abolitionism, women's suffrage and social reform were often held around the dining room table. The family performed theatricals using the dining room as their stage while guests watched from the adjoining parlor.

The parlor was a formal room with arched niches built by Bronson to display busts of his favorite philosophers, Socrates and Plato. On May 23, 1860, Anna married John Bridge Pratt in this room.

The youngest daughter, May, was a talented artist. Her bedroom contains sketches of angelic, mythological and biblical figures on the woodwork and doors. In Louisa's room, May painted a panel of calla lilies as well as an owl on the fireplace. Copies of Turner landscapes by May adorn various rooms in Orchard House.

In 1868, Louisa May wrote her beloved classic novel Little Women in her room on a special "shelf desk" built by her father. Set within the house, its characters are based on members of her family, with the plot loosely based on the family's earlier years and events that transpired at The Wayside. Also written in the house were Bronson Alcott's

Ralph Waldo Emerson (1865; published 1882), Tablets (1868), Concord Days (1872), and Table Talk (1877).

On the grounds to the west of the house is a structure designed and built by Bronson Alcott originally known as "Hillside Chapel" and later as "The Concord School of Philosophy". Operating from 1879 to 1888, the School was one of the first highly successful adult education centers in the country.

In 1877, Louisa May Alcott bought the a home on Main Street for her sister Anna. After Mrs. Alcott's death in the same year, Louisa and her father moved into the home as well. Orchard House was then sold to long-time family friend William Torrey Harris in 1884.

The Orchard House Today

Orchard House is open for guided tours daily, with the exceptions of Easter, Thanksgiving, Christmas, and January 1 and 2. An admission fee is charged.

The exterior looks much as it did in the Alcotts' day. Care has been taken to keep extensive structural preservation work invisible. All of the furnishings are original to the mid-nineteenth century, about 75% belonged to the Alcott family, and the rooms look very much as they did when the Alcotts were in residence.

The Hillside Chapel.

The dining room contains family china, portraits of the family members, and paintings by May along with period furnishings. The parlor is decorated with period wallpaper and a patterned reproduction carpet while family portraits and watercolors by May adorn the walls. Abigail May's bread board, mortar and pestle, tin spice chest and wooden bowls are displayed on the hutch table in the kitchen. Other original kitchen features include a laundry drying rack designed by Bronson, and a soapstone sink bought by Louisa. The study is furnished with Bronson's library table, chair and desk. The parent's bedroom contains many of Abigail May's possessions, including photographs, furniture, and hand made quilts.

Orchard House has continued the tradition of Mr. Alcott's Concord School of Philosophy by hosting "The Summer Conversational Series" since 1977, and has recently added a "Teacher Institute" component. The Hillside Chapel is also used for youth programs, poetry readings, historical reenactments, and other special events.

Fruita, Utah

Fruita is the best-known settlement in Capitol Reef National Park in Wayne County, Utah, United States. It is located at the confluence of Fremont River and Sulphur Creek.

History

Fruita was established in 1880 by a group of Mormons led by Nels Johnson, under the name "Junction." The town became known as Fruita in 1902 or 1904. In 1900, Fruita was named The Eden of Wayne County for its large orchards. Fruita was abandoned in 1955 when the National Park Service purchased the town to be included in Capitol Reef National Park.

Today few buildings remain, except for the restored schoolhouse and the Gifford house and barn. The orchards remain, now under the ownership of the National Park Service, and have about 2,500 trees. The orchards are preserved by the NPS as a "historic landscape" and a small crew takes care of them by pruning, irrigating, replanting, and spraying them.

The one-room schoolhouse was built and opened in 1896. The few students were instructed mainly in reading, writing, and arithmetic, but when the teachers were capable, they also studied other subjects such as history or geography. The room was also used for balls and religious services. It was renovated in 1966 by the National Park Service.

Fruita is currently the heart and administrative center of Capitol Reef National Park.

Fabaceae

This article is about Fabaceae s.l. (or Leguminosae), as defined by the APG System. For Fabaceae s.s. (or Papilionaceae), as defined by less modern systems.

The Fabaceae, Leguminosae or Papilionaceae, commonly known as the legume, pea, or bean family, are a large and economically important family of flowering plants. It includes trees, shrubs, and perennial or annual herbaceous plants, which are easily recognized by their fruit (legume) and their compound, stipulated leaves. The family is widely distributed, and is the third-largest land plant family in terms of number of species, behind only the Orchidaceae and Asteraceae, with about 751 genera and some 19,000 known species . The five largest of the genera are Astragalus (over 3,000 species), Acacia (over 1000 species), Indigofera (around 700 species), Crotalaria (around 700 species) and Mimosa (around 500 species), which constitute about a quarter of all legume species. The ca. 19,000 known legume species amount to about 7% of flowering plant species. Fabaceae is the most common family found in tropical rainforests and in dry forests in the Americas and Africa.

Recent molecular and morphological evidence supports the fact that the Fabaceae is a single monophyletic family. This point of view has been supported not only by the degree of interrelation shown by different groups within the family compared with that found among the Leguminosae and their closest relations, but also by all the recent phylogenetic studies based on DNA sequences. These studies confirm that the Fabaceae are a monophyletic group that is closely related to the Polygalaceae, Surianaceae and Quillajaceae families and that they belong to the order Fabales.

Along with the cereals, some fruits and tropical roots a number of Leguminosae have been a staple human food for millennia and their use is closely related to human evolution.

A number are important agricultural and food plants, including Glycine max (soybean), Phaseolus (beans), Pisum sativum (pea), Cicer arietinum (chickpeas), Medicago sativa (alfalfa), Arachis hypogaea (peanut), Lathyrus odoratus (sweet pea), Ceratonia siliqua (carob), and Glycyrrhiza glabra (liquorice). A number of species are also weedy pests in different parts of the world, including: Cytisus scoparius (broom), Robinia pseudoacacia (black locust), Ulex europaeus (gorse), Pueraria lobata (kudzu), and a number of Lupinus species.

Description

The fruit of Gymnocladus dioicus.

Fabaceae range in habit from giant trees (like Koompassia excelsa) to small annual herbs, with the majority being herbaceous perennials. Plants have indeterminate inflorescences, which are sometimes reduced to a single flower. The flowers have a short hypanthium and a single carpel with a short gynophore, and after fertilization produce fruits that are legumes.

Growth Habit

The Leguminosae have a wide variety of growth forms including trees, shrubs or herbaceous plants or even vines or lianas. The herbaceous plants can be annuals, biennials or perennials, without basal or terminal leaf aggregations. They are upright plants, epiphytes or vines. The latter support themselves by means of shoots that twist around a support or through cauline or foliar tendrils. Plants can be heliophytes, mesophytes or xerophytes.

Leaves

The leaves are usually alternate and compound. Most often they are even- or odd-pinnately compound (e.g. Caragana and Robinia respectively), often trifoliate (e.g. Trifolium, Medicago) and rarely palmately compound (e.g. Lupinus), in the Mimosoideae and the Caesalpinioideae commonly bipinnate (e.g. Acacia, Mimosa). They always have stipules, which can be leaf-like (e.g. Pisum), thorn-like (e.g. Robinia) or be rather inconspicuous. Leaf margins are entire or, occasionally, serrate. Both the leaves and the leaflets often have wrinkled pulvini to permit nastic movements. In some species, leaflets have evolved into tendrils (e.g. Vicia).

Many species have leaves with structures that attract ants that protect the plant from herbivore insects (a form of mutualism). Extrafloral nectaries are common among the Mimosoideae and the Caesalpinioideae, and are also found in some Faboideae (e.g. Vicia sativa). In some Acacia, the modified hollow stipules are inhabited by ants and are known as domatia.

Roots

Many Fabaceae host bacteria in their roots within structures called root nodules. These bacteria, known as rhizobia, have the ability to take nitrogen gas (N_2) out of the air and convert it to a form of nitrogen that is usable to the host plant (NO_3^- or NH_3). This process is called nitrogen fixation. The legume, acting as a host, and rhizobia, acting as a provider of usable nitrate, form a symbiotic relationship.

Flowers

The flowers often have five generally fused sepals and five free petals. They are generally hermaphrodite, and have a short hypanthium, usually cup shaped. There are normally ten stamens and one elongated superior ovary, with a curved style. They are usually arranged in indeterminate inflorescences. Fabaceae are typically entomophilous plants (i.e. they are pollinated by insects), and the flowers are usually showy to attract pollinators.

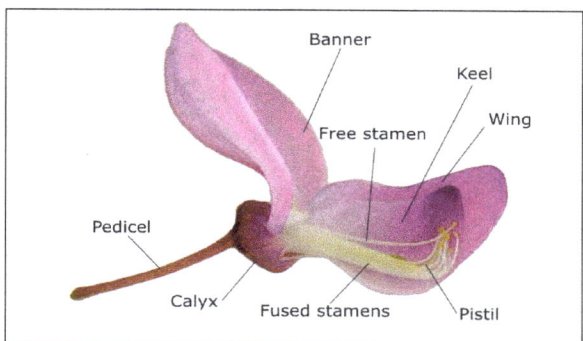

A flower of Wisteria sinensis, Faboideae. Two petals have been removed to show stamens and pistil.

In the Caesalpinioideae, the flowers are often zygomorphic, as in Cercis, or nearly symmetrical with five equal petals in Bauhinia. The upper petal is the innermost one, unlike in the Faboideae. Some species, like some in the genus Senna, have asymmetric flowers, with one of the lower petals larger than the opposing one, and the style bent to one side. The calyx, corolla, or stamens can be showy in this group.

In the Mimosoideae, the flowers are actinomorphic and arranged in globose inflorescences. The petals are small and the stamens, which can be more than just 10, have long, coloured filaments, which are the showiest part of the flower. All of the flowers in an inflorescence open at once.

In the Faboideae, the flowers are zygomorphic, and have a specialized structure. The upper petal, called the banner, is large and envelops the rest of the petals in bud, often reflexing when the flower blooms. The two adjacent petals, the wings, surround the two bottom petals. The two bottom petals are fused together at the apex (remaining free at the base), forming a boat-like structure called the keel. The stamens are always ten in number, and their filaments can be fused in various configurations, often in a group of nine stamens plus one separate stamen. Various genes in the CYCLOIDEA (CYC)/ DICHOTOMA (DICH) family are expressed in the upper (also called dorsal or adaxial) petal; in some species, such as Cadia, these genes are expressed throughout the flower, producing a radially symmetrical flower.

Fruit

The ovary most typically develops into a legume. A legume is a simple dry fruit that usually dehisces (opens along a seam) on two sides. A common name for this type of fruit is a "pod", although that can also be applied to a few other fruit types. A few species have evolved samarae, loments, follicles, indehiscent legumes, achenes, drupes, and berries from the basic legume fruit.

Legume of Vicia angustifolia.

Physiology and Biochemistry

The Leguminosae are rarely cyanogenic, however, where they are, the cyanogenic compounds are derived from tyrosine, phenylalanine or leucine. They frequently contain alkaloids. Proanthocyanidins can be present either as cyanidin or delphinidine or both at the same time. Flavonoids such as kaempferol, quercitin and myricetin are often present. Ellagic acid has never been found in any of the genera or species analysed. Sugars are transported within the plants in the form of sucrose. C3 photosynthesis has been found in a wide variety of genera. The family has also evolved a unique chemistry. Pterocarpans are a class of molecules (derivatives of isoflavonoids) found only in the Fabaceae.

Ecology

Distribution and Habitat

The Fabaceae have an essentially worldwide distribution, being found everywhere except Antarctica and the high arctic. The trees are often found in tropical regions, while the herbaceous plants and shrubs are predominant outside the tropics.

Biological Nitrogen Fixation

Biological nitrogen fixation (BNF, performed by the organisms called diazotrophs) is a very old process that probably originated in the Archean eon when the primitive atmosphere lacked oxygen. It is only carried out by Euryarchaeota and just 6 of the more than 50 phyla of bacteria. Some of these lineages co-evolved together with the flowering plants establishing the molecular basis of a mutually beneficial symbiotic relationship. BNF is carried out in nodules that are mainly located in the root cortex, although they are occasionally located in the stem as in Sesbania rostrata. The spermatophytes that co-evolved with actinorhizal diazotrophs (Frankia) or with rhizobia to establish their symbiotic relationship belong to 11 families contained within the Rosidae clade (as established by the gene molecular phylogeny of rbcL, a gene coding for part of the RuBisCO enzyme in the chloroplast). This grouping indicates that the predisposition for forming nodules probably only arose once in flowering plants and that it can be considered as an ancestral characteristic that has been conserved or lost in certain lineages. However, such a wide distribution of families and genera within this lineage indicates that nodulation had multiple origins. Of the 10 families within the Rosidae, 8 have nodules formed by actinomyces (Betulaceae, Casuarinaceae, Coriariaceae, Datiscaceae, Elaeagnaceae, Myricaceae, Rhamnaceae and Rosaceae), and the two remaining families, Ulmaceae and Fabaceae have nodules formed by rhizobia.

Roots of Vicia with white root nodules visible.

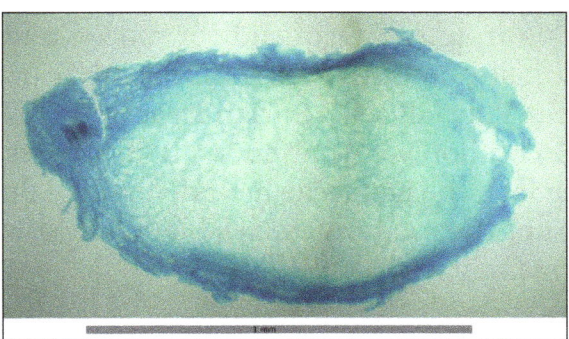

Cross-section through a root nodule of Vicia observed through a microscope.

The rhizobia and their hosts must be able to recognize each other for nodule formation to commence. Rhizobia are specific to particular host species although a rhizobia species may often infect more than one host species. This means that one plant species may be infected by more than one species of bacteria. For example, nodules in Acacia senegal can contain seven species of rhizobia belonging to three different genera. The most distinctive characteristics that allow rhizobia to be distinguished apart are the rapidity of their growth and the type of root nodule that they form with their host. Root nodules can be classified as being either indeterminate, cylindrical and often branched, and determinate, spherical with prominent lenticels. Indeterminate nodules are characteristic of legumes from temperate climates, while determinate nodules are commonly found in species from tropical or subtropical climates.

Nodule formation is common throughout the leguminosae, it is found in the majority of its members that only form an association with rhizobia, which in turn form an exclusive symbiosis with the leguminosae (with the exception of Parasponia, the only genus of the 18 Ulmaceae genera that is capable of forming nodules). Nodule formation is present in all the leguminosae sub-families, although it is less common in the Caesalpinioideae. All types of nodule formation are present in the sub-family Papilionoideae: indeterminate (with the meristem retained), determinate (without meristem) and the type included in Aeschynomene. The latter two are thought to be the most modern and specialised type of nodule as they are only present in some lines of the Papilionoideae sub-family. Even though nodule formation is common in the two monophyletic sub-families Papilionoideae and Mimosoideae they also contain species that do not form nodules. The presence or absence of nodule-forming species within the three sub-families indicates that nodule formation has arisen several times during the evolution of the leguminosae and that this ability has been lost in some lineages. For example, within the genus Acacia, a member of the Mimosoideae, A. pentagona does not form nodules, while other species of the same genus readily form nodules, as is the case for Acacia senegal, which forms both rapidly and slow growing rhizobial nodules.

Evolution, Phylogeny and Taxonomy

Evolution

The order Fabales contains around 7.3% of eudicot species and the greatest part of this diversity is contained in just one of the four families that order contains: Fabaceae. This clade also includes the Polygalaceae, Surianaceae and Quillajaceae families and its origins date back 94 to 89 million years, although it started its diversification some 79 to 74 million years ago. In fact, the Fabaceae have diversified during the early tertiary to become a ubiquitous part of the modern earth's biota, along with many other families belonging to the flowering plants.

The Fabaceae have an abundant and diverse fossil record, especially for the Tertiary period. Fossils of flowers, fruit, leaves, wood and pollen from this period have been found in numerous locations. The earliest fossils that can be definitively assigned to the Fabaceae appeared in the late Palaeocene (approximately 56 million years ago). Representatives of the 3 sub-families traditionally recognised as being members of the Fabaceae – Cesalpinioideae, Papilionoideae and Mimosoideae — as well as members of the large clades within these sub-families – such as the genistoides – have been found in periods a little later, starting between 55 and 50 million years ago. In fact, a wide variety of taxa representing the main lineages in the Fabaceae have been found in the fossil record dating from the middle to the late Eocene, suggesting that the majority of the modern Fabaceae groups were already present and that a broad diversification occurred during this period. Therefore, the Fabaceae started their diversification approximately 60 million years ago and the most important clades separated some 50 million years ago. The age of the main Cesalpinioideae clades have been estimated as between 56 and 34 million years and the basal group of the Mimosoideae as 44 ± 2.6 million years. The division between Mimosoideae and Faboideae is dated as occurring between 59 and 34 million years ago and the basal group of the Faboideae as 58.6 ± 0.2 million years ago. It has been possible to date the divergence of some of the groups within the Faboideae, even though diversification within each genus was relatively recent. For instance, Astragalus separated from the Oxytropis some 16 to 12 million years ago. In addition, the separation of the aneuploid species of Neoastragalus started 4 million years ago. Inga, another genus of the Papilionoideae with approximately 350 species, seems to have diverged in the last 2 million years.

It has been suggested, based on fossil and phylogenetic evidence, that legumes originally evolved in arid and/or semi-arid regions along the Tethys seaway during the Palaeogene Period. However, others contend that Africa (or even the Americas) cannot yet be ruled out as the origin of the family.

The current hypothesis about the evolution of the genes needed for nodulation is that they were recruited from other pathways after a polyploidy event. Several different pathways have been implicated as donating duplicated genes to the pathways need for nodulation. The main donors to the pathway were the genes associated with the arbuscular mycorrhiza symbiosis genes, the pollen tube formation genes and the haemoglobin genes. One of the main genes shown to be shared between the arbuscular mycorrhiza pathway and the nodulation pathway is SYMRK and it is involved in the plant-bacterial recognition. The pollen tube growth is similar to the infection thread development in that infection threads grow in a polar manner that is similar to a pollen tubes polar growth towards the ovules. Both pathways include the same type of enzymes, pectin-degrading cell wall enzymes. The enzymes needed to reduce nitrogen, nitrogenases, require a substantial input of ATP but at the same time are sensitive to free oxygen. To meet the requirements of this paradoxical situation, the plants express a type of haemoglobin called leghaemoglobin that is believed to be recruited after a duplication event. These three genetic pathways are believed to be part of a gene duplication event then recruited to work in nodulation.

Phylogeny and Taxonomy

Phylogeny

The phylogeny of the legumes has been the object of many studies by research groups from around the world. These studies have used morphology, DNA data (the chloroplast intron trnL, the chloroplast genes rbcL and matK, or the ribosomal spacers ITS) and cladistic analysis in order to investigate the relationships between the family's different lineages. The studies have confirmed that the traditional sub-families Mimosoideae and Papilionoideae are each monophyletic but both are nested within the paraphyletic sub-family Caesalpinioideae. All the different approaches have yielded similar results regarding the relationships between the family's main clades, as shown in the cladogram below.

Taxonomy

The Fabaceae are placed in the order Fabales according to most taxonomic systems, including the APG III system. The family includes three subfamilies:

- Mimosoideae: 80 genera and 3,200 species. Mostly tropical and warm temperate Asia and America. Mimosa, Acacia.

- Caesalpinioideae: 170 genera and 2,000 species, cosmopolitan. Caesalpinia, Senna, Bauhinia, Amherstia.

- Faboideae (Papilionoideae): 470 genera and 14,000 species, cosmopolitan. Astragalus, Lupinus.

These three subfamilies have been alternatively treated at the family level, as in the Cronquist and Dahlgren systems. However, this choice has not been supported by late 20th and early 21st century evidence, which has shown the Caesalpinioideae to be paraphyletic and the Fabaceae sensu lato to be monophyletic. While the Mimosoideae and the Faboideae are largely monophyletic, the Caesalpinioideae appear to be paraphyletic and the tribe Cercideae is probably sister to the rest of the family. Moreover, there are a number of genera whose placement into the Caesalpinioideae is not always agreed on (e.g. Dimorphandra).

Genera

The 730 genera included in this family can be viewed on the following three pages:

- List of Mimosoideae genera.

- List of Caesalpinioideae genera.

- List of Faboideae genera.

Economic and Cultural Importance

Legumes are economically and culturally important plants due to their extraordinary diversity and abundance, the wide variety of edible vegetables they represent and due to the variety of uses they can be put to: in horticulture and agriculture, as a food, for the compounds they contain that have medicinal uses and for the oil and fats they contain that have a variety of uses.

Food and Forage

The history of legumes is tied in closely with that of human civilization, appearing early in Asia, the Americas (the common bean, several varieties) and Europe (broad beans) by 6,000 BCE, where they became a staple, essential as a source of protein.

Their ability to fix atmospheric nitrogen reduces fertilizer costs for farmers and gardeners who grow legumes, and means that legumes can be used in a crop rotation to replenish soil that has been depleted of nitrogen. Legume seeds and foliage have a comparatively higher protein content than non-legume materials, due to the additional nitrogen that legumes receive through the process. Some legume species perform hydraulic lift, which makes them ideal for intercropping.

Farmed legumes can belong to numerous classes, including forage, grain, blooms, pharmaceutical/industrial, fallow/green manure and timber species, with most commercially farmed species filling two or more roles simultaneously.

There are of two broad types of forage legumes. Some, like alfalfa, clover, vetch, and Arachis, are sown in pasture and grazed by livestock. Other forage legumes such as Leucaena or Albizia are woody shrub or tree species that are either broken down by livestock or regularly cut by humans to provide stock feed.

Grain legumes are cultivated for their seeds, and are also called pulses. The seeds are used for human and animal consumption or for the production of oils for industrial uses. Grain legumes include both herbaceous plants like beans, lentils, lupins, peas and peanuts. and trees such as carob, mesquite and tamarind.

Bloom legume species include species such as lupin, which are farmed commercially for their blooms as well as being popular in gardens worldwide. Laburnum, Robinia, Gleditsia, Acacia, Mimosa, and Delonix are ornamental trees and shrubs.

Industrial farmed legumes include Indigofera, cultivated for the production of indigo, Acacia, for gum arabic, and Derris, for the insecticide action of rotenone, a compound it produces.

Fallow or green manure legume species are cultivated to be tilled back into the soil to exploit the high nitrogen levels found in most legumes. Numerous legumes are farmed for this purpose, including Leucaena, Cyamopsis and Sesbania.

Various legume species are farmed for timber production worldwide, including numerous Acacia species, Dalbergia species, and Castanospermum australe.

Melliferous plants offer nectar to bees and other insects to encourage them to carry pollen from the flowers of one plant to others thereby ensuring pollination.A number of legume species are good nectar providers such as alfalfa, white clover, sweet clover and various Prosopis species. Many plants in the Fabaceae family are an important source of pollen for the bumblebee species Bombus hortorum. This bee species is especially fond of one species in particular; Trifolium pratense, also known as red clover, is a popular food source in the diet of Bombus hortorum.

Industrial Uses

Natural Gums

Natural gums are vegetable exudates that are released as the result of damage to the plant such as that resulting from the attack of an insect or a natural or artificial cut. These exudates contain heterogeneous polysaccharides formed of different sugars and usually containing uronic acids. They form viscous colloidal solutions. There are different species that produce gums. The most important of these species belong to the leguminosae. They are widely used in the pharmaceutical, cosmetic, food and textile sectors. They also have interesting therapeutic properties; for example gum arabic is antitussive and anti-inflammatory. The most well known gums are tragacanth (Astragalus gummifer), gum arabic (Acacia senegal) and guar gum (Cyamopsis tetragonoloba).

Dyes

Indigo colorant.

The species used to produce dyes include the following: Logwood Haematoxylon campechianum; a large spiny tree that can grow up to 15 m tall. Its cork is thin and soft and its wood is hard. The heartwood is used to produce dyes that are red and purple. The histological stain called haematoxylin is produced from this species. Brazilwood tree (Caesalpinia echinata) is similar to the previous tree but smaller and with red or purple flowers. The wood is also used to produce a red or purple dye. The Madras thorn (Pithecallobium dulce) is another spiny tree native to Latin America, it grows up to 4 m high and has yellow or green flowers that grow in florets. Its fruit is reddish and is used to produce a yellow dye. Indigo dye is extracted from the True indigo plant Indigofera tinctoria that is native to Asia. In Central and South America dyes are produced from two species related to this species, indigo from Indigofera suffruticosa and Natal indigo from Indigofera arrecta.yellow dye is extracted from Butea monosperma commonly called as flame of the forest.

Ornamentals

The Cockspur Coral Tree Erythrina crista-galli is one of many leguminosae used as ornamental plants. In addition, it is the National Flower of Argentina and Uruguay.

Legumes have been used as ornamental plants throughout the world for many centuries. Their vast diversity of heights, shapes, foliage and flower colour means that this family is commonly used in the design and planting of everything from small gardens to large parks. The following is a list of the main ornamental legume species, listed by sub-family.

- Subfamily Caesalpinioideae: Bauhinia forficata, Caesalpinia gilliesii, Caesalpinia spinosa, Ceratonia siliqua, Cercis siliquastrum, Gleditsia triacanthos, Gymnocladus dioica, Parkinsonia aculeata, Senna multiglandulosa.

- Subfamily Mimosoideae: Acacia caven, Acacia cultriformis, Acacia dealbata, Acacia karroo, Acacia longifolia, Acacia melanoxylon, Acacia paradoxa, Acacia retinodes, Acacia saligna, Acacia verticillata, Acacia visco, Albizzia julibrissin, Calliandra tweediei, Paraserianthes lophantha, Prosopis chilensis.

- Subfamily Faboideae: Clianthus puniceus, Citysus scoparius, Erythrina crista-galli, Erythrina falcata, Laburnum anagyroides, Lotus peliorhynchus, Lupinus arboreus, Lupinus polyphyllus, Otholobium glandulosum, Retama monosperma, Robinia hispida, Robinia luxurians, Robinia pseudoacacia, Sophora japonica, Sophora macnabiana, Sophora macrocarpa, Spartium junceum, Teline monspessulana, Tipuana tipu, Wisteria sinensis.

Emblematic Leguminosae

- The Cockspur Coral Tree (Erythrina crista-galli), is the National Flower of Argentina and Uruguay.

- The Elephant ear tree (Enterolobium cyclocarpum) is the national tree of Costa Rica, by Executive Order of 31 August 1959.

- The Brazilwood tree (Caesalpinia echinata) has been the national tree of Brazil since 1978.

- The Golden wattle Acacia pycnantha is Australia's national flower.

- The Hong Kong Orchid tree Bauhinia blakeana is the national flower of Hong Kong.

Brassicaceae

Brassicaceae or Cruciferae is a medium-sized and economically important family of flowering plants commonly known as the mustards, the crucifers, or the cabbage family.

The name Brassicaceae is derived from the included genus Brassica. The alternative older name, Cruciferae, meaning "cross-bearing", describes the four petals of mustard flowers, which resemble a cross. Cruciferae is one of eight plant family names without the suffix -aceae that are authorized alternative names (according to ICBN Art. 18.5 and 18.6 Vienna Code).

The family contains 372 genera and 4060 accepted species. The largest genera are Draba (440 species), Erysimum (261 species), Lepidium (234 species), Cardamine (233 species), and Alyssum (207 species).

The family contains the cruciferous vegetables, including species such as Brassica oleracea (e.g., broccoli, cabbage, cauliflower, kale, collards), Brassica rapa (turnip, Chinese cabbage, etc.), Brassica napus (rapeseed, etc.), Raphanus sativus (common radish), Armoracia rusticana (horseradish), Matthiola (stock) and the model organism Arabidopsis thaliana (thale cress).

Pieris rapae and other butterflies of the family Pieridae are some of the best-known pests of Brassicaceae species planted as commercial crops.

Taxonomy

The family is included in the Brassicales according to the APG system. Older systems (e.g., Arthur Cronquist's) placed them into the Capparales, a now-defunct order that had a similar definition.

This family comprises about 365 genera and 3200 species all over the world; 94 species of 38 genera are found in Nepal. The plants are mostly herbs. A close relationship has long been acknowledged between the Brassicaceae and the caper family, Capparaceae, in part because members of both groups produce glucosinolate (mustard oil) compounds. The Capparaceae as traditionally circumscribed were paraphyletic with respect to Brassicaceae, with Cleome and several related genera being more closely related to the Brassicaceae than to other Capparaceae. The APG II system, therefore, has merged the two families under the name Brassicaceae. Other classifications have continued to recognize the Capparaceae, but with a more restricted circumscription, either including Cleome and its relatives in the Brassicaceae or recognizing them in the segregate family Cleomaceae. The APG III system has recently adopted this last solution, but this may change as a consensus arises on this point. This article deals with Brassicaceae sensu stricto, i.e. treating the Cleomaceae and Capparaceae as segregated families.

Description

Aubrieta deltoidea (commonly known as purple rock cress) is a perennial wild flower used in gardening for its ornamental large inflorescence.

The family consists mostly of herbaceous plants with annual, biennial, or perennial lifespans. However, around the Mediterranean, they include also a dozen woody shrubs 1-3 m tall, e.g. in northern Africa (Zilla spinosa and Ptilotrichum spinosum), in the Dalmatian islands (Dendralyssum and Cramboxylon), and chiefly in Canarias with some woody cruciferous genera: Dendrosinapis, Descurainia, Parolinia, Stanleya, etc.

The leaves are alternate (rarely opposite), sometimes organized in basal rosettes; in rare shrubby crucifers of Mediterranean their leaves are mostly in terminal rosettes, and may be coriaceous and evergreen. They are very often pinnately incised and do not have stipules.

The structure of the flowers is extremely uniform throughout the family. They have four free saccate sepals and four clawed free petals, staggered. They can be disymmetric or slightly zygomorphic, with a typical cross-like arrangement (hence the name Cruciferae). They have six stamens, four of which are longer (as long as the petals) and are arranged in a cross like the petals and the other two are shorter (tetradynamous flower). The pistil is made up of two fused carpels and the style is very short, with two lobes. The ovary is superior. The flowers form ebracteate racemose inflorescences, often apically corymb-like.

Pollination occurs by entomogamy; nectar is produced at the base of the stamens and stored on the sepals.

Siliquae of Cardamine impatiens.

The fruit is a peculiar kind of capsule named siliqua (plural siliquae). It opens by two valves, which are the modified carpels, leaving the seeds attached to a framework made up of the placenta and tissue from the junction between the valves (replum). Often, an indehiscent beak occurs at the top of the style and one or more seeds may be borne there. Where a siliqua is less than three times as long as it is broad, it is usually termed a silicula. The siliqua may break apart at constrictions occurring between the segments of the seeds, thus forming a sort of loment (e.g., Raphanus), it may eject the seeds explosively (e.g., Cardamine) or may be evolved in a sort of samara (e.g., Isatis). The fruit is often the most important diagnostic character for plants in this family. Most members share a suite of glucosinolate compounds that have a typical pungent odour usually associated with cole crops.

Uses

The importance of this family for food crops has led to its selective breeding throughout history. Some examples of cruciferous food plants are the cabbage, broccoli, cauliflower, turnip, rapeseed, mustard, radish, horseradish, cress, wasabi, and watercress.

Lunaria annua with ripe seed pods.

Smelowskia americana is endemic to the midlatitude mountains of western North America.

Cruciferous vegetables

Matthiola (stock), Cheiranthus, Lobularia, and Iberis (candytufts) are appreciated for their flowers. Lunaria (honesty) is cultivated for the decorative value of the translucent replum of the round silicula that remains on the dried stems after dehiscence.

Capsella bursa-pastoris, Lepidium, and many Cardamine species are common weeds.

Isatis tinctoria (woad) was used in the past to produce the colour indigo.

Arabidopsis thaliana is a very important model organism in the study of the flowering plants (Angiospermae).

Cucurbitaceae

The Cucurbitaceae, also called cucurbits and the gourd family, are a plant family consisting of ca 965 species in around 95 genera, the most important of which are:

- Cucurbita – squash, pumpkin, zucchini, some gourds.

- Lagenaria – mostly inedible gourds.

- Citrullus – watermelon (C. lanatus, C. colocynthis) and others.

- Cucumis – cucumber (C. sativus), various melons.

- Luffa – the common name is also luffa, sometimes spelled loofah (when fully ripened, two species of this fibrous fruit are the source of the loofah scrubbing sponge).

The plants in this family are grown around the tropics and in temperate areas, where those with edible fruits were among the earliest cultivated plants both in the Old and New Worlds. The Cucurbitaceae family ranks among the highest of plant families for number and percentage of species used as human food.

Pumpkins and squashes displayed in a show competition.

The Cucurbitaceae consist of 98 proposed genera with 975 species, mainly in regions tropical and subtropical. All species are sensitive to frost. Most of the plants in this family are annual vines, but some are woody lianas, thorny shrubs, or trees (Dendrosicyos). Many species have large, yellow or white flowers. The stems are hairy and pentangular. Tendrils are present at 90° to the leaf petioles at nodes. Leaves are exstipulate alternate simple palmately lobed or palmately compound. The flowers are unisexual, with male and female flowers on different plants (dioecious) or on the same plant (monoecious). The female flowers have inferior ovaries. The fruit is often a kind of modified berry called a pepo.

Fossil History

One of the oldest fossil records so far is Cucurbitaciphyllum lobatum from the Paleo-cene epoch, found at Shirley Canal, Montana. It was described for the first time in 1924 by Knowlton. The fossil leaf is palmate, trilobed with rounded lobal sinuses and an entire or serrate margin. It has a leaf pattern similar to the members of the genera Kedrotis, Melothria and Zehneria.

Classification

A selection of cucurbits of the South Korean Genebank in Suwon.

Cucurbits on display at the Real Jardín Botánico de Madrid, with the title "Variedades de calabaza".

The following is a classification given by Charles Jeffrey in 1990. However, a 2011 study based on genetics does not support this taxonomy with two subfamilies and eight tribes, but rather delineates 15 tribes, five of them new, consisting of 95 genera rather than Jeffrey's 121.

Subfamily Zanonioideae (small striate pollen grains):

- Tribe Zanonieae:

    ○ Subtribe Fevilleinae: Fevillea.

    ○ Subtribe Zanoniinae: Alsomitra Zanonia Siolmatra Gerrardanthus Zygosicyos Xerosicyos Neoalsomitra.

    ○ Subtribe Gomphogyninae: Hemsleya Gomphogyne Gynostemma.

- ○ Subtribe Actinostemmatinae: Bolbostemma Actinostemma.

- ○ Subtribe Sicydiinae: Sicydium Chalema Pteropepon Pseudosicydium Cyclantheropsis.

Subfamily Cucurbitoideae (styles united into a single column):

- Tribe Melothrieae:

    - ○ Subtribe Dendrosicyinae: Kedrostis Dendrosicyos Corallocarpus Ibervillea Tumamoca Halosicyos Ceratosanthes Doyerea Trochomeriopsis Seyrigia Dieterlea Cucurbitella Apodanthera Guraniopsis Melothrianthus Wilbrandia.

    - ○ Subtribe Guraniinae: Helmontia Psiguria Gurania.

    - ○ Subtribe Cucumerinae: Melancium Cucumeropsis Posadaea Melothria Muellarargia Zehneria Cucumis (including: Mukia, Dicaelospermum, Cucumella, Oreosyce, and Myrmecosicyos).

    - ○ Subtribe Trochomeriinae: Solena Trochomeria Dactyliandra Ctenolepis.

- Tribe Schizopeponeae: Schizopepon.

- Tribe Joliffieae:

    - ○ Subtribe Thladianthinae: Indofevillea Siraitia Thladiantha Momordica.

    - ○ Subtribe Telfairiinae: Telfairia.

- Tribe Trichosantheae:

    - ○ Subtribe Hodgsoniinae: Hodgsonia.

    - ○ Subtribe Ampelosicyinae: Ampelosicyos Peponium.

    - ○ Subtribe Trichosanthinae: Gymnopetalum Trichosanthes Tricyclandra.

    - ○ Subtribe Herpetosperminae: Cephalopentandra Biswarea Herpetospermum Edgaria.

- Tribe Benincaseae:

    - ○ Subtribe Benincasinae: Cogniauxia Ruthalicia Lagenaria Benincasa Praecitrullus Citrullus Acanthosicyos Eureiandra Bambekea Nothoalsomitra Coccinia Diplocyclos Raphidiocystis Lemurosicyos Zombitsia Ecballium Bryonia.

    - ○ Subtribe Luffinae: Luffa.

- Tribe Cucurbiteae (pantoporate, spiny pollen): Cucurbita Sicana Tecunumania

Calycophysum Peponopsis Anacaona Polyclathra Schizocarpum Penelopeia Cionosicyos Cayaponia Selysia Abobra.

- Tribe Sicyeae (trichomatous nectary, four- to 10-colporate pollen grains):

  ◦ Subtribe Cyclantherinae: Hanburia Echinopepon Marah Echinocystis Vaseyanthus Brandegea Apatzingania Cremastopus Elateriopsis Pseudocyclanthera Cyclanthera Rytidostylis.

  ◦ Subtribe Sicyinae: Sicyos Sicyosperma Parasicyos Microsechium Sechium Sechiopsis Pterosicyos.

- Incertae sedis: Odosicyos.

Alphabetical list of genera: Abobra Acanthosicyos Actinostemma Alsomitra Ampelosycios Anacaona Apatzingania Apodanthera Bambekea Benincasa Biswarea Bolbostemma Brandegea Bryonia Calycophysum Cayaponia Cephalopentandra Ceratosanthes Chalema Cionosicyos Citrullus Coccinia Cogniauxia Corallocarpus Cremastopus Ctenolepis Cucumella Cucumeropsis Cucumis Cucurbita Cucurbitella Cyclanthera Dactyliandra Dendrosicyos Dicaelospermum Dieterlea Diplocyclos Doyerea Ecballium Echinocystis Echinopepon Edgaria Elateriopsis Eureiandra Fevillea Gerrardanthus Gomphogyne Gurania Guraniopsis Gymnopetalum Gynostemma Halosicyos Hanburia Helmontia Hemsleya Herpetospermum Hodgsonia Ibervillea Indofevillea Kedrostis Lagenaria Lemurosicyos Luffa Marah Melancium Melothria Melothrianthus Microsechium Momordica Muellerargia Mukia Myrmecosicyos Neoalsomitra Nothoalsomitra Odosicyos Oreosyce Parasicyos Penelopeia Peponium Peponopsis Polyclathra Posadaea Praecitrullus Pseudocyclanthera Pseudosicydium Psiguria Pteropepon Pterosicyos Raphidiocystis Ruthalicia Rytidostylis Schizocarpum Schizopepon Sechiopsis Sechium Selysia Seyrigia Sicana Sicydium Sicyos Sicyosperma Siolmatra Siraitia Solena Tecunumania Telfairia Thladiantha Trichosanthes Tricyclandra Trochomeria Trochomeriopsis Tumacoca Vaseyanthus Wilbrandia Xerosicyos Zanonia Zehneria Zombitsia Zygosicyos.

Round melons and elongate adzhur melons in Kursi church mosaic, Israel, near the Sea of Galilee.

Images of Cucurbits in Byzantine Mosaics from Israel

Six cucurbit crops are represented in 23 Byzantine-era mosaics from Israel, these being round melons (Cucumis melo), watermelons (Citrullus lanatus), sponge gourds (Luffa aegyptiaca), snake melons (faqqous, Cucumis melo Flexuosus Group), adzhur melons (Cucumis melo Adzhur Group), and bottle gourds (Lagenaria siceraria). Cucurbits are represented in 23 of the 134 mosaics containing images of crop plants, a surprisingly high frequency of 17%. Several of the cucurbit images have not been found elsewhere, suggesting a diverse and highly developed local horticulture of cucurbits in Israel during the Byzantine era. Representations of mature sponge gourds are found in widespread localities, suggestive of the high value accorded to cleanliness and hygiene.

References

- Singh, Gurcharan (2004). Plants Systematics: An Integrated Approach. Science Publishers. p. 83. ISBN 1-57808-351-6

- Schlegel, Rolf H J (January 1, 2003). Encyclopedic Dictionary of Plant Breeding and Related Subjects. Haworth Press. p. 177. ISBN 1-56022-950-0

- Mauseth, James D. (April 1, 2003). Botany: An Introduction to Plant Biology. Jones and Bartlett. pp. 271–272. ISBN 0-7637-2134-4

- McGee, Harold (November 16, 2004). On Food and Cooking: The Science and Lore of the Kitchen. Simon & Schuster. pp. 247–248. ISBN 0-684-80001-2

- Mauseth, James D. (2003). Botany: an introduction to plant biology. Boston: Jones and Bartlett Publishers. p. 258. ISBN 978-0-7637-2134-3

- Luther Burbank. Practical Orchard Plans and Methods: How to Begin and Carry on the Work. The Minerva Group. ISBN 1-4147-0141-1

- Cothran, James R. (November 1, 2003). Gardens and Historic Plants of the Antebellum South. University of South Carolina Press. p. 221. ISBN 1-57003-501-6

- Farrell, Kenneth T. (November 1, 1999). Spices, Condiments and Seasonings. Springer. pp. 17–19. ISBN 0-8342-1337-0

- Adams, Denise Wiles (February 1, 2004). Restoring American Gardens: An Encyclopedia of Heirloom Ornamental Plants, 1640–1940. Timber Press. ISBN 0-88192-619-1

- Heiser, Charles B. (April 1, 2003). Weeds in My Garden: Observations on Some Misunderstood Plants. Timber Press. pp. 93–95. ISBN 0-88192-562-4

Chapter 4

Preservation Techniques

Preservation slows down the activity of disease-causing bacteria in food. There are various methods of preserving vegetables and fruits such as freezing, drying and canning. These methods of preservation of vegetables and fruits as well as their storage have been thoroughly discussed in this chapter.

Freezing of Fruits and Vegetables

Freezing is one of the oldest and most widely used methods of food preservation, which allows preservation of taste, texture, and nutritional value in foods better than any other method. The freezing process is a combination of the beneficial effects of low temperatures at which microorganisms cannot grow, chemical reactions are reduced, and cellular metabolic reactions are delayed.

The Importance of Freezing as a Preservation Method

Freezing preservation retains the quality of agricultural products over long storage periods. As a method of long-term preservation for fruits and vegetables, freezing is generally regarded as superior to canning and dehydration, with respect to retention in sensory attributes and nutritive properties. The safety and nutrition quality of frozen products are emphasized when high quality raw materials are used, good manufacturing practices are employed in the preservation process, and the products are kept in accordance with specified temperatures.

The Need for Freezing and Frozen Storage

Freezing has been successfully employed for the long-term preservation of many foods, providing a significantly extended shelf life. The process involves lowering the product temperature generally to -18 °C or below. The physical state of food material is changed when energy is removed by cooling below freezing temperature. The extreme cold simply retards the growth of microorganisms and slows down the chemical changes that affect quality or cause food to spoil.

Competing with new technologies of minimal processing of foods, industrial freezing is the most satisfactory method for preserving quality during long storage periods. When compared in terms of energy use, cost, and product quality, freezing requires the shortest processing time. Any other conventional method of preservation focused on fruits and vegetables, including dehydration and canning, requires less energy when compared with energy consumption in the freezing process and storage. However, when the overall cost is estimated, freezing costs can be kept as low (or lower) as any other method of food preservation.

Current Status of Frozen Food Industry in U.S. and other Countries

The frozen food market is one of the largest and most dynamic sectors of the food industry. In spite of considerable competition between the frozen food industry and other sectors, extensive quantities of frozen foods are being consumed all over the world. The industry has recently grown to a value of over US$ 75 billion in the U.S. and Europe combined. This number has reached US$ 27.3 billion in 2001 for total retail sales of frozen foods in the U.S. alone. In Europe, based on U.S. currency, frozen food consumption also reached 11.1 million tons in 13 countries in the year 2000. Table represents the division of frozen food industry in terms of annual sales in 2001.

Advantages of Freezing Technology in Developing Countries

Developed countries, mostly the U.S., dominate the international trade of fruits and vegetables. The U.S. is ranked number one as both importer and exporter, accounting for the highest percent of fresh produce in world trade. However, many developing countries still lead in the export of fresh exotic fruits and vegetables to developed countries.

For developing countries, the application of freezing preservation is favorable with several main considerations. From a technical point of view, the freezing process is one of the most convenient and easiest of food preservation methods, compared with other commercial preservation techniques. The availability of different types of equipment for several different food products results in a flexible process in which degradation of initial food quality is minimal with proper application procedures. the high capital investment of the freezing industry usually plays an important role in terms of economic feasibility of the process in developing countries. As for cost distribution, the freezing process and storage in terms of energy consumption constitute approximately 10 percent of the total cost. Depending on the government regulations, especially in developing countries, energy cost for producers can be subsidized by means of lowering the unit price or reducing the tax percentage in order to enhance production. Therefore, in determining the economical convenience of the process, the cost related to energy consumption (according to energy tariffs) should be considered. Electricity prices for some countries are given in Table.

Table: Frozen food industry in terms of annual sales in 2001.

Food items	Sales US$ (million)	% Change vs. 2000
Total Frozen Food Sales	26 600	6.1
Baked Goods	1 400	9.0
Breakfast Foods	1 050	4.1
Novelties	1 900	10.5
Ice Cream	4 500	5.7
Frozen Dessert/Fruit/Toppings	786	5.4
Juices/Drinks	827	-9.7
Vegetables	2 900	4.3

Increasing Consumer Demand in Developing Countries due to Modernization

The proportion of fresh food preserved by freezing is highly related to the degree of economic development in a society. As countries become wealthier, their demand for high-valued commodities increases, primarily due to the effect of income on the consumption of high-valued commodities in developing countries. The commodities preserved by freezing are usually the most perishable ones, which also have the highest price. Therefore, the demand for these commodities is less in developing areas. Besides, the need for adequate technology for freezing process is the major drawback of developing countries in competing with industrialized countries. The frozen food industry requires accompanying developments and facilities for transporting, storing, and marketing their products from the processing plant to the consumer. Thus, a large amount of capital investment is needed for these types of facilities. For developing countries, especially in rural or semi-rural areas, the frozen food industry has therefore not been developed significantly compared to other countries.

In recent years, due to the changing consumer profile, the frozen food industry has changed significantly. The major trend in consumer behavior documented over the last half century has been the increase in the number of working women and the decline in the family size. These two factors resulted in a reduction in time spent preparing food. The entry of more women into the workforce also led to improvements in kitchen appliances and increased the variability of ready-to-eat or frozen foods available in the market. Besides, the increased usage of microwave ovens, affecting food habits in general and the frozen food market in particular, as well as allowing rapid preparation of meals and greater flexibility in meal preparation. The frozen food industry is now only limited by imagination, an output of which increases continuously to supply the increasing demand for frozen products and variability.

Table: Unit electricity prices for industry[1] (U.S. Dollars per Kilowatt-hour).

Country	1999	2000	2001	2002
Argentina	n.a.	0.075	0.069	n.a.
Belgium	0.056	0.048	n.a.	n.a.
Bolivia	n.a.	0.062	0.069	n.a.
Chile	n.a.	0.052	0.056	n.a.
Chinese Taipei (Taiwan)	0.058	0.061	0.056	n.a.
Colombia	n.a.	0.052	0.042	n.a.
Costa Rica	n.a.	0.068	0.076	n.a.
Cuba	n.a.	0.080	0.078	n.a.
Ecuador	n.a.	0.036	0.061	n.a.
El Salvador	n.a.	0.111	0.110	n.a.
Finland	0.046	0.039	0.038	0.043
Germany	0.057	0.041	0.044	n.a.
Greece	0.050	0.042	0.043	0.046
Guyana	n.a.	0.082	0.080	n.a.
Hungary	0.055	0.049	0.051	0.060
India	0.081	0.080	n.a.	n.a.

Ireland	0.057	0.049	0.060	0.075
Italy	0.086	0.089	n.a.	n.a.
Korea (Korea, South)	0.056	0.062	0.057	n.a.
Mexico	0.042	0.051	0.053	n.a.
Netherlands	0.061	0.057	0.059	n.a.
New Zealand	0.030	0.030	0.028	0.033
Nicaragua	n.a.	0.117	0.115	n.a.
Paraguay	n.a.	0.032	0.036	n.a.
Peru	n.a.	0.056	0.057	n.a.
Poland	0.037	0.037	0.045	0.049
Portugal	0.078	0.067	0.066	0.068
Russia	0.012	0.011	n.a.	n.a.
South Africa	0.017	0.017	0.013	n.a.
Spain	0.049	0.043	0.041	n.a.
Switzerland	0.090	0.069	0.069	0.073
Turkey	0.079	0.080	0.079	0.094
United Kingdom	0.064	0.055	0.048	n.a.
United States [2]	0.044	0.046	0.050	0.048
Uruguay	n.a.	0.064	0.070	n.a.

n.a. = Not Available.

[1] Energy end-use prices including taxes converted using exchange rates. [2] Electricity prices in the United States, including income taxes, environmental charges, and other charges.

Market Share of Frozen Fruits and Vegetables

Today in modern society, frozen fruits and vegetables constitute a large and important food group among other frozen food products. The historical development of commercial freezing systems designed for special food commodities helped shape the frozen food market. Technological innovations as early as 1869 led to the commercial development and marketing of some frozen foods. Early products saw limited distribution through retail establishments due to insufficient supply of mechanical refrigeration. Retail distribution of frozen foods gained importance with the development of commercially frozen vegetables in 1929.

The frozen vegetable industry mostly grew after the development of scientific methods for blanching and processing in the 1940s. Only after the achievement of success in stopping enzymatic degradation, did frozen vegetables gain a strong retail and institutional appeal. Today, market studies indicate that considering overall consumption of frozen foods, frozen vegetables constitute a very significant proportion of world frozen-food categories (excluding ice cream) in Austria, Denmark, Finland, France, Germany, Italy, Netherlands, Norway, Sweden, Switzerland, UK, and the USA. The division of frozen vegetables in terms of annual sales in 2001 is shown in Table.

Commercialization history of frozen fruits is older than frozen vegetables. The commercial freezing of small fruits and berries began in the eastern part of the U.S. in about 1905. The main advantage of freezing preservation of fruits is the extended usage of frozen fruits during off-season.

Additionally, frozen fruits can be transported to remote markets that could not be accessed with fresh fruit. Also, freezing preservation makes year-round further processing of fruit products possible, such as jams, juice, and syrups from frozen whole fruit, slices, or pulps. The preservation of fruits by freezing has clearly become one the most important preservation methods.

Future Trends in Freezing Technology

The frozen food industry is highly based in modern science and technology. Starting with the first historical development in freezing preservation of foods, today, a combination of several factors influences the commercialization and usage of freezing technology. The future growth of frozen foods will mostly be affected by economic and technological factors. Growth in population, personal incomes, and relative cost of other forms of foods, changes in tastes and preferences, and technological advances in freezing methods are some of the factors concerned with the future of freezing technology.

Population growth and increasing demand for food has generated the need for commercial production of food commodities in large-scale operations. Thus, availability of proper equipment suitable for continuous processing would be valuable for freezing preservation methods. In addition depending on personal incomes, relative cost of frozen products is one of the most important of economic factors. Producing the highest quality at the lowest cost possible is highly dependent on the technology used. As a result, developments in freezing technology in recent years have mostly been characterized by the improvements in mechanical handling and process control to increase freezing rate and reduce cost.

Today an increasing demand for frozen foods already exits and further expansion of the industry is primarily dependent on the ability of food processors to develop higher qualities in both process techniques and products. Improvements can only be achieved by focusing on new technologies and investigating poorly understood factors that influence the quality of frozen food products. Improvements in new and convenient forms of foods, as well as more information on relative cost and nutritive values of frozen foods, will contribute toward continued growth of the industry.

Table: Frozen vegetables in terms of annual sales in 2001.

Vegetables	Sales US$ (million)	% Change vs. 2000
Broccoli	184	4.4
Com/Corn on the Cob	312	3.5
Green Beans	115	6.0
Mixed Vegetables	450	7.2
Peas	207	3.9
Potatoes	1 070	4.4

General Recommendations on the Freezing Process

Freezing is a widely used method of food preservation based on several advantages in terms of retention of food quality and ease of process. Beginning with the earliest history of freezing, the technology has been highly affected over the years by the developments and improvements in freezing techniques. In order to understand and handle the concepts associated with freezing of foods, it is necessary to examine the fundamental factors governing the freezing process.

Freezing Technology

Freezing has long been used as a method of preservation, and history reveals it was mostly shaped by the technological developments in the process. A small quantity of ice produced without using a "natural cold" in 1755 was regarded as the first milestone in the freezing process. Firstly, ice-salt systems were used to preserve fish and later on, by the late 1800's, freezing was introduced into large-scale operations as a method of commercial preservation. Meat, fish, and butter, the main products preserved in this early example, were frozen in storage chambers and handled as bulk commodities.

In the following years, scientists and researchers continuously worked to achieve success with commercial freezing trials on several food commodities. Among these commodities, fruits were one of the most important since freezing during the peak growing season had the advantage of preserving fruit for later processing into jams, jellies, ice cream, pies, and other bakery foods. Although commercial freezing of small fruits and berries first began around 1905 in the eastern part of the United States, the commercial freezing of vegetables is much more recent. Starting from 1917, only private firms conducted trials on freezing vegetables, but achieving good quality in frozen vegetables was not possible without pre-treatments due to the enzymatic deterioration. The necessity of blanching to inactivate enzymes before freezing was concluded by several researchers to avoid deterioration and off-flavours caused by enzymatic degradation.

The modern freezing industry began in 1928 with the development of double-belt contact freezers by a technologist named Clarence Birdseye. After the revolution in the quick freezing process and equipment, the industry became more flexible, especially with the usage of multi-plate freezers. The earlier methods achieved successful freezing of fish and poultry, however with the new quick freezing system, packaged foods could be frozen between two metal belts as they moved through a freezing tunnel. This improvement was a great advantage in the commercial large-scale freezing of fruits and vegetables. Furthermore, quick-freezing of consumer-size packages helped frozen vegetables to be accepted rapidly in late 1930s.

Today, freezing is the only large-scale method that bridges the seasons, as well as variations in supply and demand of raw materials such as meat, fish, butter, fruits, and vegetables. Besides, it makes possible movement of large quantities of food over geographical distances. It is important to control the freezing process, including the pre-freezing preparation and post-freezing storage of the product, in order to achieve high-quality products. Therefore, the theory of the freezing process and the parameters involved should be understood clearly.

Freezing Process

The freezing process mainly consists of thermodynamic and kinetic factors, which can dominate each other at a particular stage in the freezing process. Major thermal events are accompanied by reduction in heat content of the material during the freezing process as is shown in Figure. The material to be frozen first cools down to the temperature at which nucleation starts. Before ice can form, a nucleus, or a seed, is required upon which the crystal can grow; the process of producing this seed is defined as nucleation. Once the first crystal appears in the solution, a phase change occurs from liquid to solid with further crystal growth. Therefore, nucleation serves as the initial process of freezing, and can be considered as the critical step that results in a complete phase change.

Freezing Point of Foods

Freezing point is defined as the temperature at which the first ice crystal appears and the liquid at that temperature is in equilibrium with the solid. If the freezing point of pure water is considered, this temperature will correspond to 0 °C (273°K). However, when food systems are frozen, the process becomes more complex due to the existence of both free and bound water. Bound water does not freeze even at very low temperatures. Unfreezable water contains soluble solids, which cause a decrease in the freezing point of water lower than 0 °C. During the freezing process, the concentration of soluble solids increases in the unfrozen water, resulting in a variation in freezing temperature. Therefore, the temperature at which the first ice crystal appears is commonly regarded as the initial freezing temperature. There are empirical equations in literature that can calculate the initial freezing temperature of certain foods as a function of their moisture content.

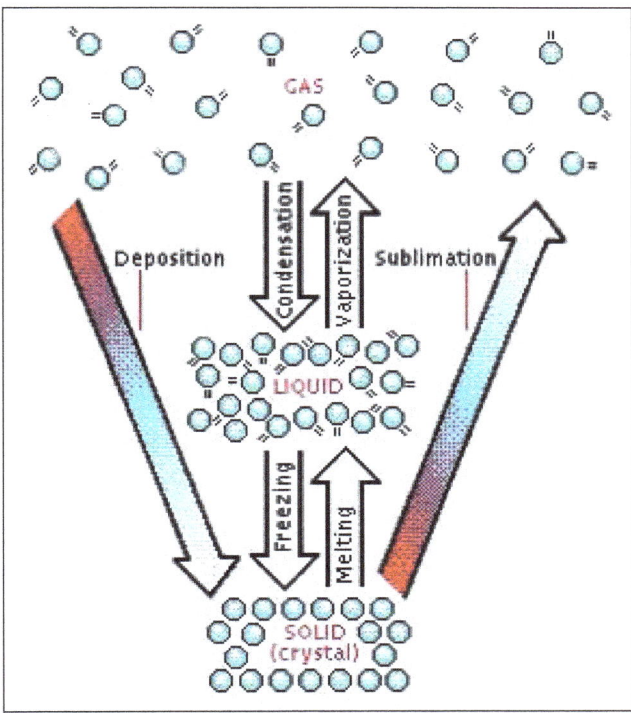

A schematic illustration of overall freezing process.

There are several methods of food freezing, and depending on the method used, the quality of the frozen food may vary. However, regardless of the method chosen, the main principle behind all freezing processes is the same in terms of process parameters. The International Institute of Refrigeration (IIR) has provided definitions to establish a basis for the freezing process. According to their definition, the freezing process is basically divided into three stages based on major temperature changes in a particular location in the product, as shown in Figures and for pure water and food respectively.

Beginning with the prefreezing stage, the food is subjected to the freezing process until the appearance of the first crystal. If the material frozen is pure water, the freezing temperature will be 0 °C and, up to this temperature, there will be a sub cooling until the ice formation begins. In the case of foods during this stage, the temperature decreases to below freezing temperature and, with the formation of the first ice crystal, increases to freezing temperature. The second stage is the freezing period; a phase change occurs, transforming water into ice. For pure water, temperature at this stage is constant; however, it decreases slightly in foods, due to the increasing concentration of solutes in the unfrozen water portion. The last stage starts when the product temperature reaches the point where most freezable water has been converted to ice, and ends when the temperature is reduced to storage temperature.

The freezing time and freezing rate are the most important parameters in designing freezing systems. The quality of the frozen product is mostly affected by the rate of freezing, while time of freezing is calculated according to the rate of freezing. For industrial applications, they are the most essential parameters in the process when comparing different types of freezing systems and equipment.

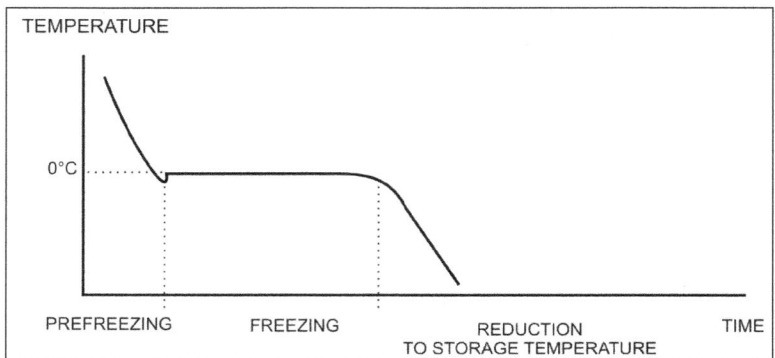

Practical definition of the freezing process for pure water.

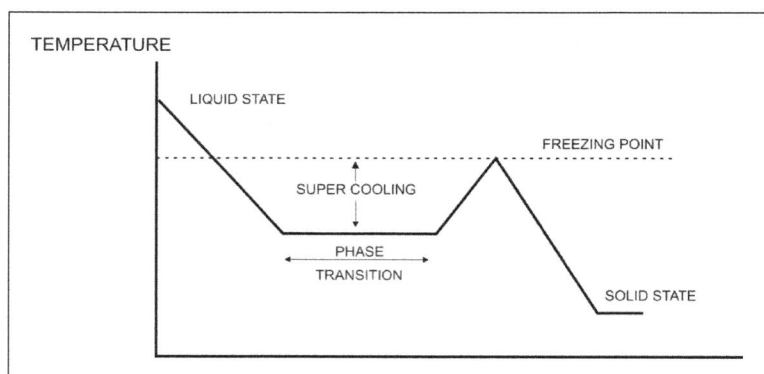

Practical definition of the freezing process for foods.

Freezing Time

Again, freezing time is one of the most important parameters in the freezing process, defined as time required to lower product temperature from its initial temperature to a given temperature at its thermal center. Since the temperature distribution within the product varies during freezing process, the thermal center is generally taken as reference. Thus, when the geometrical center of the product reaches the given final temperature, this ensures the average product temperature has been reduced to a storage value. Freezing time depends on several factors, including the initial and final temperatures of the product and the quantity of heat removed, as well as dimensions (especially thickness) and shape of product, heat transfer process, and temperature. The International Institute of Refrigeration defines various factors of freezing time in relation to both the product frozen and freezing equipment. The most important are:

- Dimensions and shape of product, particularly thickness.

- Initial and final temperatures.

- Temperature of refrigerating medium.

- Surface heat transfer coefficient of product.

- Change in enthalpy.

- Thermal conductivity of product.

Calculation of freezing time in food systems is difficult in comparison to pure systems since the freezing temperature changes continuously during the process. Using a simplified approach, time elapsed between initial freezing until when the entire product is frozen can be regarded as the freezing time. Plank's equation ($t_F = \dfrac{\rho\lambda_1}{T_F - T_e}\left[\dfrac{e^2 R}{k} + \dfrac{eP}{h}\right]$) is commonly used to estimate freezing time, however due to assumptions involved in the calculation it is only useful for obtaining an approximation of freezing time. The derivation of the equation starts with the assumption the product being frozen is initially at freezing temperature. Therefore, the calculated freezing time represents only the freezing period. The equation can be further modified for different geometries including slab, cylinder, and sphere, where for each geometry, the coefficients are arranged in relation to the dimensions.

Table: Coefficients P and R of equation given below:

Geometry	P	R	Dimension
Infinite slab	1/2	1/8	thickness e
Infinite cylinder	1/4	1/16	radius r
Sphere	1/6	1/24	radius r

$$t_F = \frac{\rho\lambda_1}{T_F - T_e}\left[\frac{e^2 R}{k} + \frac{eP}{h}\right]$$

where λ_1 is the latent heat of frozen fraction, k and r are the thermal conductivity and density of the frozen layer, while h is the coefficient of heat transfer by convection to the exterior. T_f denotes the body temperature of the product when introduced into a freezer in wich the external temperature is T_e The coefficients R and P are given in table and arranged according to the geometry of the product frozen. where the letter e denotes the dimension (i.e. for infinite slab geometry, e is thickness of the slab and for infinite cylinder or sphere e is replaced by r which denotes the radius of the clylinder or sphere).

the equation of Plank assumes the food is at a freezing temperature at the beginning of the freezing process. However, the food is usually at a temperature higher than freezing temperature. The real freezing time should therefore be the sum of time calculated from the equation of Plank and the time needed for the product's surface to decrease from initial temperature to freezing temperature.

Several works have attempted to calculate real freezing time, as in one presented by Nagaoka. Nagaoka's equation ($t_F = \frac{\rho \Delta \lambda_1}{T_F - T_e}\left[\frac{Re^2}{k} + \frac{PI}{h}\right]\left[1 + 0.008(T_i - T_F)\right]$) calculates the amount of heat elimination required to decrease a product's temperature from initial temperature to freezing temperature, as well as the amount of heat released during the phase change and the amount of heat eliminated to reach freezing temperature.

$$t_F = \frac{\rho \Delta \lambda_1}{T_F - T_e}\left[\frac{Re^2}{k} + \frac{PI}{h}\right]\left[1 + 0.008(T_i - T_F)\right]$$

where T_i is the temperature of the food at the initiation of freezing, DH is the difference between the enthalpy of the food at initial temperature and end of freezing. Re and Pl are the dimensionless numbers, while k and h are the thermal conductivity and the coefficient of heat transfer, respectively.

For calculating freezing time of products with irregular shape, a common property of most food products - especially fruits and vegetables - a dimensionless factor has been employed in equations.

Freezing Rate

The freezing rate (°C/h) for a product or package is defined as the ratio of difference between initial and final temperature of product to freezing time. At a particular location within the product, a local freezing rate can be defined as the ratio of the difference between the initial temperature and desired temperature to the time elapsed in reaching the given final temperature. The quality of frozen products is largely dependent on the rate of freezing. Generally, rapid freezing results in better quality frozen products when compared with slow freezing. If freezing is instantaneous, there will be more locations within the food where crystallization begins. In contrast, if freezing is slow, the crystal growth will be slower with few nucleation sites resulting in larger ice crystals. Large ice crystals are known to cause mechanical damage to cell walls in addition to cell dehydration. Thus, the rate of freezing for plant tissues is extremely important due to the effect of freezing rate on the size of ice crystals, cell hydration, and damage to cell walls. The figure shows a general behavior of the dynamics curve of freezing preservation.

Rapid freezing is advantageous for freezing of many foods, however some products are susceptible to cracking when exposed to extremely low temperature for long periods.

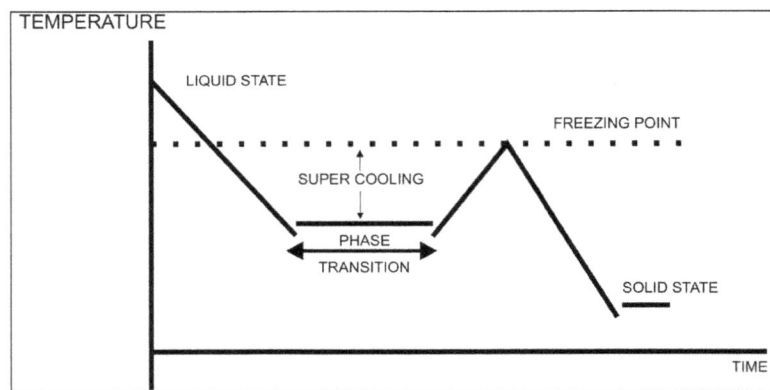

Freezing preservation dynamics curve.

Energy Requirements

For fruits and vegetables, the amount of energy required for freezing is calculated based on the enthalpy change and the amount of product to be frozen. The following equation is for calculation of refrigeration requirements for fruits and vegetables.

$$\Delta H = \left[1 - \frac{X_{SNJ}}{100}\right]\Delta H_j + 1.21\left[\frac{X_{SNJ}}{100}\right]\Delta T$$

X_{SNJ}: Percentage of the product solids different from juice (Dry matter fraction of the juice).

ΔH_j: Enthalpy change during freezing of the juice fraction.

ΔT: Temperature difference between initial and final temperature of the product.

Refrigeration

Refrigeration is defined as the elimination of heat from a material at a temperature higher than the temperature of its surroundings. The mechanism of refrigeration is a part of the freezing process and freezing storage involved in the thermodynamic aspects of freezing. According to the second law of thermodynamics, heat only flows from higher to lower temperatures. Therefore, in order to raise the heat from a lower to a higher temperature level, expenditure of work is needed. The aim of industrial refrigeration processes is to eliminate heat from low temperature points towards points with higher temperature. For this reason, either closed mechanical refrigeration cycles in which refrigeration fluids circulate, or open cryogenic systems with liquid nitrogen (LIN) or carbon dioxide (CO_2), are commonly used by the food industry.

The main elements in a closed mechanical refrigeration system are the condenser, compressor, evaporator, and the expansion valve. The refrigerants hydro chlorofluorocarbon (HCFC) and ammonia are examples of the refrigerants circulated in these types of mechanical refrigeration systems. A simple scheme for the closed mechanical refrigeration system is shown in Figure.

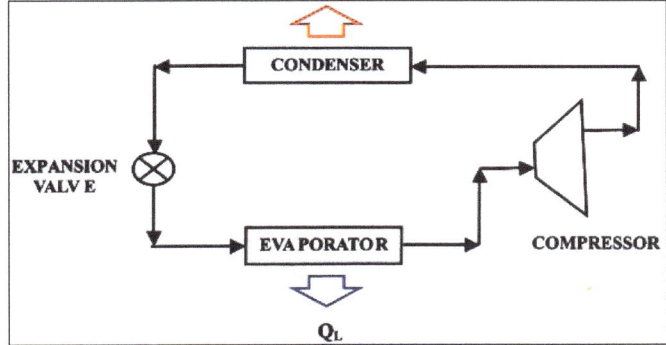

A simple scheme for a one-stage closed mechanical refrigeration system.

Starting at the suction point of the compressor, fluid in a vapor state is compressed into the compressor where an increase in temperature and pressure takes place. The fluid then flows through the condenser where it decreases in energy by giving off heat and converting to a liquid state. After the phase, a change occurs inside the condenser, the fluid flows through the expansion valve where the pressure decreases to convert liquid into a form of liquid-gas mixture. Finally, the liquid-gas mixture flows through the evaporator where it is converted into a saturated vapor state and removes heat from the environment in the process of cooling. With this last stage the loop restarts again.

The other refrigeration system employed by the food industry is the cryogenic system with carbon dioxide or liquid nitrogen. The refrigerant in this system is consumed differently from the circulating fluid in closed mechanical systems.

Refrigerants

There are several refrigerants available for refrigeration systems. The selection of a proper refrigerant is based on physical, thermodynamic, and chemical properties of the fluid. Environmental considerations are also important in refrigerant selection, since leaks within the system produce deleterious effects on the atmospheric ozone layer. Some refrigerants, including halocarbons, have been banned to avoid potential hazardous effects. For industrial applications, ammonia is commonly used, while chlorofluoromethane and tetrafluoroethane are also recommended as refrigerants.

Freezing Capacity

Freezing equipment selection is based on the requirements for freezing a certain quantity of food per hour. For any type of freezer, freezing capacity (expressed in tonnes per hour) is defined as the ratio of the quantity of the product that can be loaded into the freezer to the holding time of the product in that particular freezer. The first parameter, the amount of food product loaded into the freezer, is affected by both the dimensions of the product and the mechanical constraints of the freezer. The denominator (holding time) has an important role in freezing systems and is based on the calculation of the amount of heat removed from the product per hour, which varies depending on the type of product frozen.

Freezing Systems

There is a variety of freezing systems available for freezing, and for most products, more than one type of freezer can be used. Therefore, in selecting a freezing system initially, a cost-benefit analysis should be conducted based on three important factors: economics, functionality, and feasibility. Financial considerations mainly involve capital investment and the production cost of selected equipment. Product losses during freezing operation should be included in cost estimation since generating higher cost freezers may have other benefits in terms of reducing product losses. Functional factors are mostly based on the suitability of the selected freezer for particular products. The mode of process, either in-line or batch, should be considered based on the fact that computerized systems are becoming more important for ease of handling and lowering production costs. Mechanical constraints for the freezer should also be considered since some types of freezers are not physically suitable for freezing certain products. Lastly, the feasibility of the process should be considered in terms of plant location or location of the processing area, as well as cleanability and hygienic design, and desired product quality.

These factors and initial considerations can help eliminate several choices in freezer selection, but the relative importance of factors may change depending on the process. For developing countries where the freezing application is relatively new, the cost factor becomes more important than other factors due to the decreased production rates and need for lower capital investment costs.

Freezing Equipment

The industrial equipment for freezing can be categorized in many ways, namely as equipment used for batch or in-line operation, heat transfer systems (air, contact, cryogenic), and product stability. The rate of heat transfer from the freezing medium to the product is important in defining the freezing time of the product. Therefore, the equipment selected for freezing process characterizes the rate of freezing.

Air-blast Freezers

The air blast freezer is one the oldest and commonly used freezing equipment due to its temperature stability and versatility for several product types. In general, air is used as the freezing medium in the freezing design, either as still air or forced air. Freezing is accomplished by placing the food in freezing rooms called sharp freezers. Still, air freezing is the cheapest way of freezing and has the added advantage of a constant temperature during frozen storage, which allows usage for unprocessed bulk products like beef quarters and fish. However, it is the slowest method of freezing due to the low surface heat transfer coefficient of circulating air inside the room. Freezing time in sharp freezers is largely dependent on the temperature of the freezing chamber and the type, initial temperature, and size of product. An improved version of the still air freezer is the forced air freezer, which consists of air circulation by convection inside the freezing room. However, even modification of the sharp freezer with extra refrigeration capacity and fans for increased air circulation does not help control the air flow over the products during slow freezing. A typical design for air blast freezers is shown in figure.

There are a considerable number of designs and arrangements for air blast freezers, primarily grouped in two categories depending on the mode of process, as either inline or batch. Continuous freezers are the most suitable systems for mass production of packaged products with similar freezing times, in which the product is carried through on trucks or on conveyors. The system works on a semi-batch principle when trucks are used, since they remain stationary during the process except when a new truck enters one end of the tunnel, thus moving the others along to release a finished one at the exit. The batch freezers are more flexible since a variety of products can be frozen at the same time on individual trolleys. Over-loading may be a problem for these types of freezers, thus the process requires closer supervision than continuous systems.

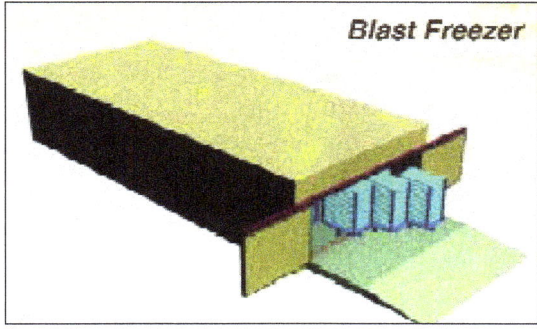

Tunnel Freezers

In tunnel freezers, the products on trays are placed in racks or trolleys and frozen with cold air circulation inside the tunnel. In order to allow air circulation, optimum space is provided between layers of trolley, which can be moved continuously in and out of the freezer manually or by forklift trucks. This freezing system is suitable for all types of products, although there are some mechanical constraints including the requirement of high manpower for handling, cleaning, and transportation of trays. A trolley for a tunnel freezer is shown in Figure.

Trolley in a tunnel freezer.

Belt Freezers

Belt freezers were first designed to provide continuous product flow with the help of a wire mesh conveyor inside the blast rooms. A poor heat transfer mechanism and the mechanical problems were solved in modern belt freezers by providing a vertical airflow to force air through the product layer. Airflow has good contact with the product only when the entire product is evenly distributed over the conveyor belt. In order to decrease required floor space, the belts can be arranged in a multi-tier belt freezer or a spiral belt freezer. Spiral belt freezers consist of a belt that can be bent laterally around a rotating drum to maximize belt surface area in a given floor space. This type of design has the advantage of eliminating product damage in transfer points, especially for products that require gentle handling. Both packed and unpacked products with long freezing times (10 min to 3 hr) can be frozen in spiral belt freezers due to the flexibility of the equipment. A typical spiral belt freezer is shown in Figure.

Fluidized Bed Freezers

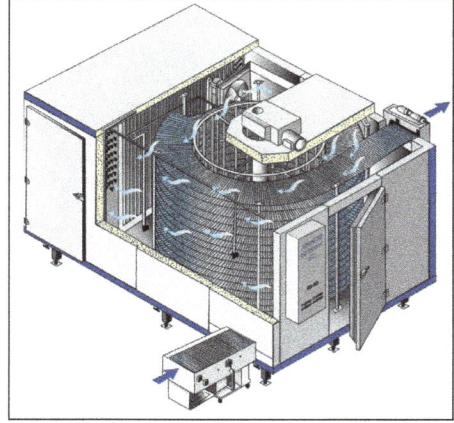

The cross-section view of a spiral belt freezer.

The fluidized bed freezer, a fairly recent modified type of air-blast freezer for particular product types, consists of a bed with a perforated bottom through which cold air is blown vertically upwards. The system relies on forced cold air from beneath the conveyor belt, causing the products to suspend or float in the cold air stream. The use of high air velocity is very effective for freezing unpacked foods, especially when they can be completely surrounded by flowing air, as in the case of fluidized bed freezers.

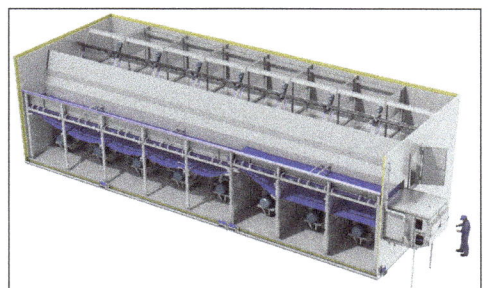
Cross-sectional view of a fluidized bed freezer.

The use of fluidization has several advantages compared with other methods of freezing since the product is individually quick frozen (IQF), which is convenient for particles with a tendency to stick together. The idea of individually quick frozen foods (IQF) started with the first technological developments aimed at quick freezing. The need for an effective means of freezing small particles with the potential for lumping during the process is the objective of IQF freezing. Small vegetables, prawns, shrimp, french-fried potatoes, diced meat, and fruits are some of the products now frozen with this technology. A typical fluidized-bed freezer is shown in Figures.

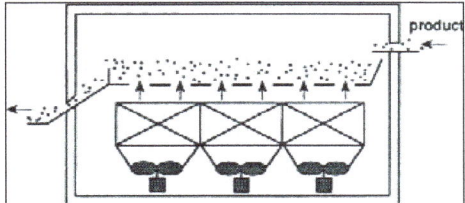
Simple working principle of a fluidized bed freezer.

Contact Freezers

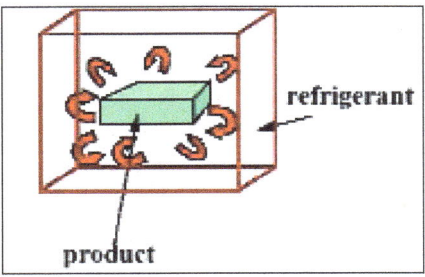
Direct contact freezer.

Contact freezing is the one of the most efficient ways of freezing in terms of heat transfer mechanism. In the process of freezing, the product can be in direct or indirect contact with the freezing medium. For direct contact freezers, the product being frozen is fully surrounded by the freezing medium, the refrigerant, maximizing the heat transfer efficiency. For indirect contact freezers, the product is indirectly exposed to the freezing medium while in contact with the belt or plate, which is in contact with the freezing medium.

Immersion Freezers

The immersion freezer consists of a tank with a cooled freezing media, such as glycol, glycerol, sodium chloride, calcium chloride, and mixtures of salt and sugar. The product is immersed in this solution or sprayed while being conveyed through the freezer, resulting in fast temperature reduction through direct heat exchange. Direct immersion of a product into a liquid refrigerant is the most rapid way of freezing since liquids have better heat conducting properties than air. The solute used in the freezing system should be safe without taste, odour, colour, or flavour, and for successful freezing, products should be greater in density than the solution. Immersion freezing systems have been commonly used for shell freezing of large particles due to the reducing ability of product dehydration when the outer layer is frozen quickly. A commonly seen problem in these freezing systems is the dilution of solution with the product, which can change the concentration and process parameters. Thus, in order to avoid product contact with the liquid refrigerant, flexible membranes can be used. A simple illustration of the immersion freezer is shown in figure.

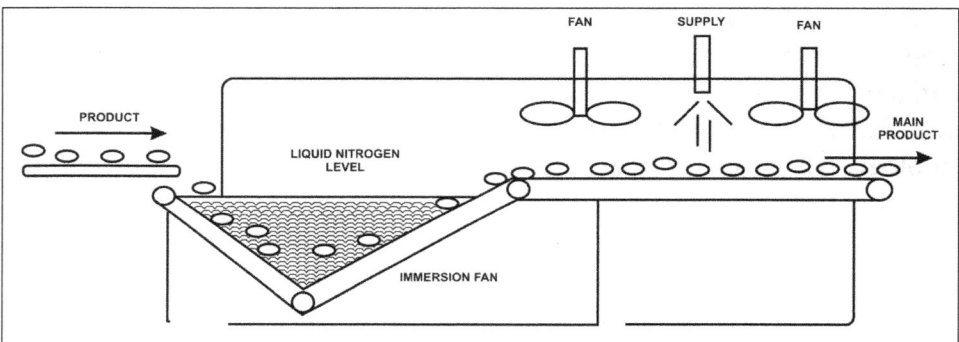
Simple illustration of a typical immersion freezer.

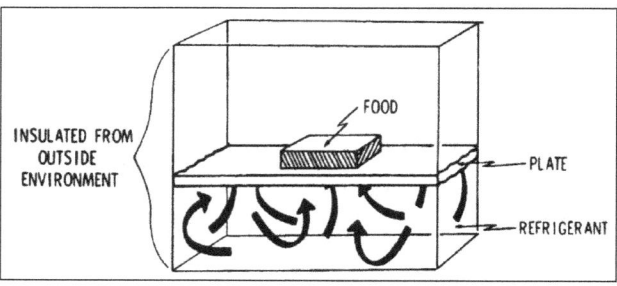
Indirect contact freezer.

Indirect Contact Freezers

In this type of freezer, materials being frozen are separated from the refrigerant by a conducting material, usually a steel plate. The mechanism of indirect contact freezer is shown in Figure. Indirect contact freezers generally provide an efficient medium for heat transfer, although the system has some limitations, especially when used for packaged foods due to resistance of package to heat transfer. Additionally, corrosive effects may occur due to interaction of metal packages with heat transfer surfaces.

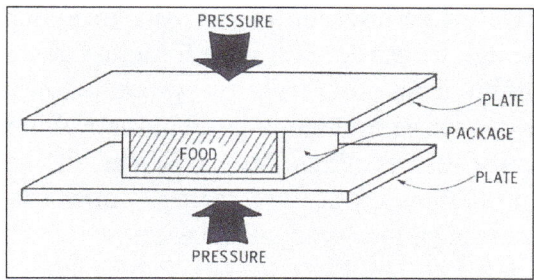

Pressure application in a plate freezer.

Plate Freezers

The most common type of contact freezer is the plate freezer. In this case, the product is pressed between hallow metal plates, either horizontally or vertically, with a refrigerant circulating inside the plates. Pressure is applied for good contact as schematically shown in Figure.

Plate freezer with a two-stage compressor and sea water condenser.

This type of freezing system is only limited to regular-shaped materials or blocks like beef patties or block-shaped packaged products. A typical plate freezer is shown in Figure.

Contact Belt Freezers

This type of freezer is designed with single-band or double-band for freezing of thin product layers as shown in Figure. The design can be either straight forward or drum. Typical products frozen in belt freezers are, fruit pulps, egg yolk, sauces and soups.

Cryogenic Freezers

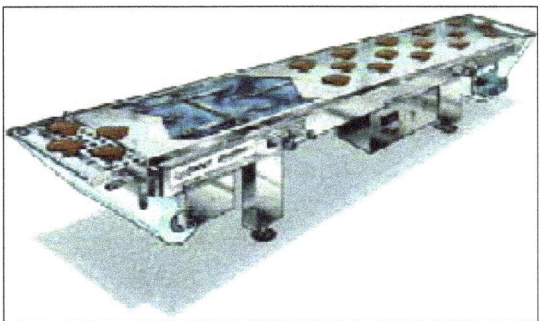

Contact belt freezer.

Cryogenic freezing is a relatively new method of freezing in which the food is exposed to an atmosphere below -60 °C through direct contact with liquefied gases such as nitrogen or carbon dioxide. This type of system differs from other freezing systems since it is not connected to a refrigeration plant; the refrigerants used are liquefied in large industrial installations and shipped to the food-freezing factory in pressure vessels. Thus, the small size and mobility of cryogenic freezers allow for flexibility in design and efficiency of the freezing application. Low initial investment and rather high operating costs are typical for cryogenic freezers.

Liquid Nitrogen Freezers

Liquid nitrogen, with a boiling temperature of -196 °C at atmospheric pressure, is a by-product of oxygen manufacture. The refrigerant is sprayed into the freezer and evaporates both on leaving the spray nozzles and on contact with the products. The system is designed in a way that the refrigerant passes in counter current to the movement of the products on the belt giving high transfer efficiency. The refrigerant consumption is in the range of 1.2-kg refrigerant per kg of the product. Typical food products used in this system are, fish fillets, seafood, fruits, berries.

Liquid Carbon Dioxide Freezers

Liquid carbon dioxide exists as either a solid or gas when stored at atmospheric pressure. When the gas is released to the atmosphere at -70 °C, half of the gas becomes dry-ice snow and the other half stays in the form of vapor. This unusual property of liquid carbon dioxide is used in a variety of freezing systems, one of which is a pre-freezing treatment before the product is exposed to nitrogen spray.

Packaging

Proper packaging of frozen food is important to protect the product from contamination and damage while in transit from the manufacturer to the consumer, as well as to preserve food value, flavour, colour, and texture. There are several factors considered in designing a suitable package for a frozen food. The package should be attractive to the consumer, protected from external contamination, and effective in terms of processing, handling, and cost. Proper selection is based on the type of package and material. There are typically three types of packaging used for frozen foods: primary, secondary, and tertiary. The primary package is in direct contact with the food and the food is kept inside the package up to the time of use. Secondary packaging is a form of multiple packaging used to handle packages together for sale. Tertiary packaging is used for bulk transportation of products.

Packaging materials should be moisture-vapor-proof to prevent evaporation, thus retaining the highest quality in frozen foods. Oxygen should also be completely evacuated from the package using a vacuum or gas-flush system to prevent migration of moisture and oxygen. Glass and rigid plastic are examples of moisture-vapor-proof packaging materials. Many packaging materials, however, are not moisture-vapor-proof, but are sufficiently moisture-vapor-resistant to retain satisfactory quality in foods. Most bags, wrapping materials, and waxed cartons used in freezing packaging are moisture-vapor-resistant. In general, the containers should be leakage free while easy to seal. Durability of the material is another important factor to consider, since the packaging material must not become brittle at low temperatures and crack.

A range of different packaging materials, mainly grouped as rigid and non-rigid containers, can be used for primary packaging. Glass, plastic, tin, and heavily waxed cardboard materials are in the rigid container group and usually used for packaging of liquid food products. Glass containers are mostly used for fruits and vegetables if they are not water-packed. Plastics are the derivatives of the oil-cracking industry. Non-rigid containers include bags and sheets made of moisture-vapor-resistant heavy aluminum foil, polyethylene or laminated papers. Bags are the most commonly used packaging materials for frozen fruits and vegetables due to their flexibility during processing and handling. They can be used with or without outer cardboard cartons to protect against tearing.

Shape and size of the container are also important factors in freezing products. Serving size may vary depending on the type of product and selection should be based on the amount of food determined for one meal. For shape of the container, freezer space must be considered since rigid containers with flat tops and bottoms stack well in the freezer, while round containers waste freezer space.

Frozen Storage and Distribution

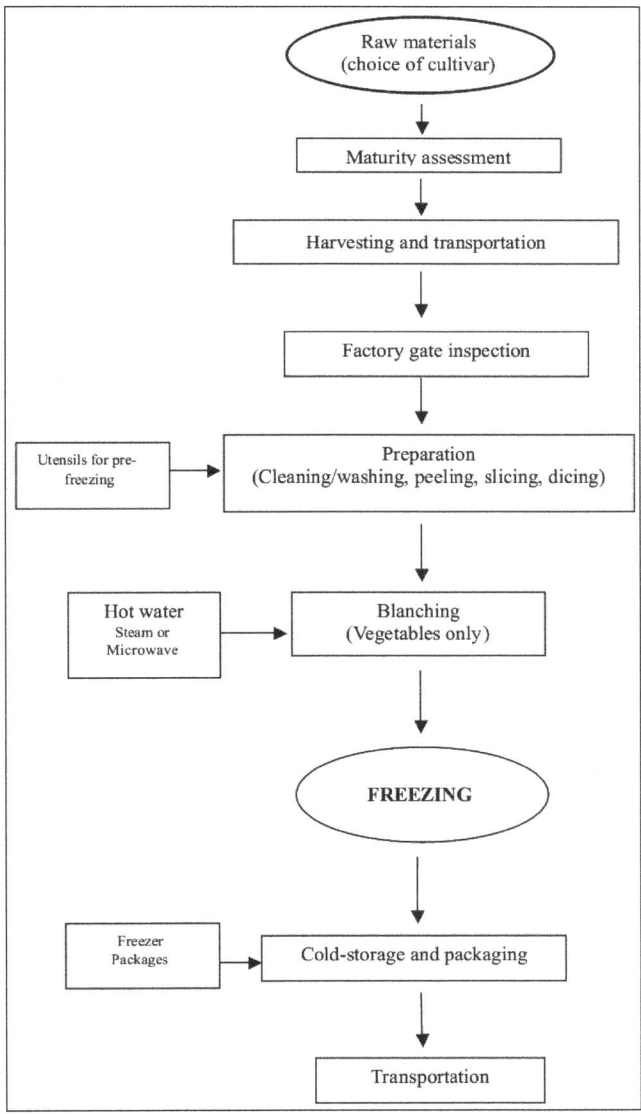

A general flow chart of frozen fruits and vegetables.

The quality of the final product depends on the history of the raw material. Using the lowest possible temperature is essential for frozen storage, transport, and distribution in achieving a high-quality product, since deteriorative processes are mainly temperature dependent. The lower the product temperature is, the slower the speed of reaction is leading to loss of quality. The temperatures of supply chains in freezing applications from the factory to the retail cabinet should be carefully monitored. The temperature regime covering the freezing process, the cold-store temperatures (£ -18 °C), distribution temperatures (£ -15 °C), and retail display (£ -12 °C) are given as legal standards.

Freezing fruits and vegetables in small and medium scale operations and its potential applications in warm climates. The preservation of fruits and vegetables by freezing is one of the most important methods for retaining high quality in agricultural products over long-term storage. In particular, the freshness qualities of raw fruits and vegetables can be retained for long periods, extending well beyond the normal season of most horticultural crops. The potential application of freezing preservation of fruits and vegetables, including tropical products, has been increasing recently in parallel with developments in developing countries.

Freezing Fruits

The effect of freezing, frozen storage, and thawing on fruit quality has been investigated over several decades. Today frozen fruits constitute a large and important food group. The quality demanded in frozen fruit products is mostly based on the intended use of the product. If the fruit is to be eaten without any further processing after thawing, texture characteristics are more important when compared to use as a raw material in other industries. In general, conventional methods of freezing tend to destroy the turgidity of living cells in fruit tissue. Different from vegetables, fruits do not have a fibrous structure that can resist this destructive effect. Additionally, fruits to be frozen are harvested in a fully ripe state and are soft in texture. On the contrary, a great number of vegetables are frozen in an immature state. Fruits have delicate flavours that are easily damaged or changed by heat, indicating they are best eaten when raw and decrease in quality with processing. In the same way, attractive colour is important for frozen fruits. Chemical treatments or additives are often used to inactivate the deteriorative enzymes in fruits. Therefore, proper processing is essential for all steps involved, from harvesting to packaging and distribution. A freezing guide for freezing fruits is shown in table.

Production and Harvesting

The characteristics of raw materials are of primary importance in determining the quality of the frozen product. These characteristics include several factors such as genetic makeup, climate of the growing area, type of fertilization, and maturity of harvest.

The ability to withstand rough handling, resistance to virus diseases, molds, uniformity in ripening, and yield are some of the important characteristics of fruits in terms of economic aspects considered in production. The use of mechanical harvesting generally causes bruising of fruits and results in a wide range of maturity levels for fruits. In contrast, hand-picking provides gentler handling and maturity sorting of fruits. However in most cases, it is non-economical compared to mechanical harvesting due to high labor cost.

As a rule, harvesting of fruits at an optimum level for commercial use is difficult. Simple tests like pressure tests are applied to determine when a fruit has reached optimum maturity for harvest.

Colour is also one of the characteristics used in determining maturity since increased maturation causes a darker colour in fruits. A combination of colour and pressure tests is a better way to assess maturity level for harvesting.

Controlled atmosphere storage is a common method of storage for some fruits prior to freezing. In principle, a controlled atmosphere high in carbon dioxide and low in oxygen content slows down the rate of respiration, which may extend shelf life of any respiring fruit during storage. Due to the fact that these fruits do not ripen appreciably after picking, most fruits are picked as near to eating-ripe maturity as possible.

Pre-process Handling and Operations

Freezing preservation of fruits can only help retain the inherent quality present initially in a product since the process does not improve the quality characteristics of raw materials. Therefore, quality level of the raw materials prior to freezing is the major consideration for successful freezing. Washing and cutting generally results in losses when applied after thawing. Thus, fruits should be prepared prior to the freezing process in terms of peeling, slicing or cutting. Freezing preservation does not require specific unit operations for cleaning, rinsing, sorting, peeling, and cutting of fruits.

Fruits that require peeling before consumption should be peeled prior to freezing. Peeling is done by scalding the fruit in hot water, steam or hot lye solutions. The effect of peeling on the quality of frozen products has been studied for several fruits, including kiwi, banana, and mango. The rate of freezing can be increased by decreasing the size of products frozen, especially for large fruits. An increase in the freezing rate results in smaller ice crystals, which decreases cellular damage in fruit tissue. Banana, tomato, mango, and kiwi are some examples of large fruits commonly cut into smaller cubes or slices prior to freezing.

The objective of blanching is to inactivate the enzymes causing detrimental changes in colour, odour, flavour, and nutritive value, but heat treatment causes loss of such characteristics in fruits. Therefore, only a few types of fruits are blanched for inactivation of enzymes prior to freezing. The loss of water-soluble minerals and vitamins during blanching should also be minimized by keeping blanching time and temperature at an optimum combination.

Effect of Ingredients

Addition of sugars is an extremely important pre-treatment for fruits prior to freezing since the treatment has the effect of excluding oxygen from the fruit, which helps to retain colour and appearance. Sugars when dissolved in solutions act by withdrawing water from cells by osmosis, resulting in very concentrated solutions inside the cells. The high concentration of solutes depresses the freezing point and therefore reduces the freezing within the cells, which inhibits excessive structural damage. Sugar syrups in the range of 30-60 percent sugar content are commonly used to cover the fruit completely, acting as a barrier to oxygen transmission and browning. Several experiments have shown the protective effect of sugar on flavour, odour, colour, and nutritive value during freezing, especially for frozen berries.

Packaging

Fruits exposed to oxygen are susceptible to oxidative degradation, resulting in browning and reduced storage life of products. Therefore, packaging of frozen fruits is based on excluding air from the fruit tissue. Replacement of oxygen with sugar solution or inert gas, consuming the oxygen by glucose-oxidase and/or the use of vacuum and oxygen-impermeable films are some of the methods currently employed for packaging frozen fruits. Plastic bags, plastic pots, paper bags, and cans are some of the most commonly used packaging materials (with or without oxygen removal) selected, based on penetration properties and thickness.

There are several types of fruit packs suitable for freezing: syrup pack, sugar pack, unsweetened pack, and tray pack and sugar replacement pack. The type of pack is usually selected according to the intended use for the fruit. Syrup-packed fruits are generally used for cooking purposes, while dry-packed and tray-packed fruits are good for serving raw in salads and garnishes.

Syrup Pack

The proportion of sugar to water used in a syrup pack depends on the sweetness of the fruit and the taste preference of the consumer. For most fruits, 40 percent sugar syrup is recommended. Lighter syrups are lower in calories and mostly desirable for mild-flavoured fruits to prevent masking the flavour, while heavier syrups may be used for very sour fruits.

Syrup is prepared by dissolving the sugar in warm water and cooling the solution down before usage. Just enough cooled syrup is used to cover the prepared fruit after it has been settled by jarring the container. In order to keep the fruit under the syrup, a small piece of crumpled waxed paper or other water resistant wrapping material is placed on top; the fruit is pressed down into the syrup before closing, then sealed and frozen.

Pectin can be used to reduce sugar content in syrups when freezing berries, cherries, and peaches. Pectin syrups are prepared by dissolving 1 box of powdered pectin with 1 cup of water. The solution is stirred and boiled for 1 minute; 1/2 cup of sugar is added and dissolved; the solution is then cooled down with the addition of cold water. Previously prepared fruit is put into a 4 to 6 quart bowl and enough pectin syrup is added to cover the fruit with a thin film. The pack is sealed and promptly frozen.

Sugar Packs

In preparing a sugar pack, sugar is first sprinkled over the fruit. Then the container is agitated gently until the juice is drawn out and the sugar is dissolved. This type of pack is generally used for soft sliced fruits such as peaches, strawberries, plums, and cherries, by using sufficient syrup to cover the fruit. Some whole fruits may also be coated with sugar prior to freezing.

Unsweetened Packs

Unsweetened packs can be prepared in several ways, either dry-packed, covered with water containing ascorbic acid, or packed in unsweetened juice. When water or juice is used in syrup and sugar packs, fruit is submerged by using a small piece of crumpled water-resistant material. Generally, unsweetened packs yield a lower quality product when compared with sugar packs, with the exception, some fruits such as raspberries, blueberries, scalded apples, gooseberries, currants, and cranberries maintain good quality without sugar.

Tray Packs

Unsweetened packs are generally prepared by using tray packs in which a single layer of prepared fruit is spread on shallow trays, frozen, and packaged in freezer bags promptly. The fruit sections remain loose without clumping together, which offers the advantage of using frozen fruit piece by piece.

Sugar Replacement Packs

Artificial sweeteners can be used instead of sugar in the form of sugar substitutes. The sweet taste of sugar can be replaced by using these kinds of sweeteners, however the beneficial effects of sugar like colour protection and thick syrup cannot be replaced. Fruits frozen with sugar substitutes will freeze harder and thaw more slowly than fruits preserved with sugar.

Freezing Vegetables

Freezing is often considered the simplest and most natural way of preservation for vegetables. Frozen vegetables and potatoes form a significant proportion of the market in terms of frozen food consumption. The quality of frozen vegetables depends on the quality of fresh products, since freezing does not improve product quality. Pre-process handling, from the time vegetables are picked until ready to eat, is one of the major concerns in quality retention.

Table: Fruit freezing guide.

Fruit	Preparation	Type of Pack
Apples	Wash, peel, and slice into antidark-ening solution - 3 tablespoons lemon juice per quart of water	Pack in 30-40% syrup, adding 1/2 teaspoon crystalline ascorbic acid per quart of syrup. Pack dry or with up to 1/2 cup sugar per quart of apple slices.
Apricots	Wash, halve, and pit. Peel and slice if desired. If apricots are not peeled, heat in boiling water for 1/2 minute to keep skins from toughening during freez-ing. Cool in cold water, drain.	Pack in 40% syrup, adding 3/4 teaspoon crystalline ascorbic acid per quart of syrup.
Avocados	Peel soft, ripe avocados. Cut in half, remove pit, mash pulp.	Add 1/8 teaspoon crystalline ascorbic acid to each quart of puree. Package in recipe-size amounts.
Berries	Select firm, fully ripe berries. Sort, wash, and drain.	Use 30% syrup pack, dry unsweetened pack, dry sugar pack, (3/4 cup sugar per quart of berries), or tray pack.
Cherries (sour or sweet)	Select well-colored, tree-ripened cherries. Stem, sort, and wash thoroughly. Drain and pit.	Pack in 30-40% syrup. Add 1/2 teaspoon ascorbic acid per quart of syrup. For pies and other cooked products, pack in dry sugar using 3/4-cup sugar per quart of fruit.
Citrus fruits, (sections or slices)	Select firm fruit, free of soft spots. Wash and peel.	Pack in 40% syrup or in fruit juice. Add 1/2 teaspoon ascorbic acid per quart of syrup or juice.

Grapes	Select firm, ripe grapes. Wash and remove stems. Leave seedless grapes whole. Cut grapes with seeds in half and remove seeds.	Pack in 20% syrup or pack without sugar. Use dry pack for halved grapes and tray pack for whole grapes.
Melons (cantaloupe, watermelon)	Select firm-fleshed, well-colored, ripe melons. Wash rinds well. Slice or cut into chunks.	Pack in 30% syrup or pack dry using no sugar. Pulp also may be crushed (except watermelon), adding 1 tablespoon sugar per quart. Freeze in recipe-size containers.

Crop Cultivar, Production and Maturity

The choice of the right cultivar and maturity before crop is harvested are the two most important factors affecting raw material quality. Raw material characteristics are usually related to the vegetable cultivar, crop production, crop maturity, harvesting practices, crop storage, transport, and factory reception.

The choice of crop cultivars is mostly based on their suitability for frozen preservation in terms of factory yield and product quality. Some of the characteristics used as selection criteria are as follows:

- Suitability for mechanical harvesting.

- Uniform maturity.

- Exceptional flavour and uniform colour and desirable texture.

- Resistance to diseases.

- High yield.

Although cultivar selection is a major factor affecting the quality of the final product, many practices in the field and factors during growth of crop can also have a significant effect on quality. Those practices include site selection for growth, nutrition of crop, and use of agricultural chemicals to control pests or diseases. The maturity assessment for harvesting is one of the most difficult parts of the production. In addition to conventional methods, new instruments and tests have been developed to predict the maturity of crops that help determining the optimum harvest time, although the maturity assessment differs according to crop variety.

Harvesting

At optimum maturity, physiological changes in several vegetables take place very rapidly. Thus, the determination of optimum harvesting time is critical (Arthey, 1993). Some vegetables such as green peas and sweet corn only have a short period during which they are of prime quality. If harvesting is delayed beyond this point, quality deteriorates and the crop may quickly become unacceptable. Most of the vegetables are subjected to bruising during harvesting.

Pre-process Handling

Vegetables at peak flavour and texture are used for freezing. Postharvest delays in handling vegetables are known to produce deterioration in flavour, texture, colour, and nutrients. Therefore, the delays between harvest and processing should be reduced to retain fresh quality prior to freezing. Cooling vegetables by cold water, air blasting, or ice will often reduce the rate of post-harvest losses sufficiently, providing extra hours of high quality retention for transporting raw material to considerable distances from the field to the processing plant.

Blanching

Blanching is the exposure of the vegetables to boiling water or steam for a brief period of time to inactivate enzymes. Practically every vegetable (except herbs and green peppers) needs to be blanched and promptly cooled prior to freezing, since heating slows or stops the enzyme action, which causes vegetables to grow and mature. After maturation, however, enzymes can cause loss in quality, flavour, colour, texture, and nutrients. If vegetables are not heated sufficiently, the enzymes will continue to be active during frozen storage and may cause the vegetables to toughen or develop off-flavours and colours. Blanching also causes wilting or softening of vegetables, making them easier to pack. It destroys some bacteria and helps remove any surface dirt.

Blanching in hot water at 70 to 105 °C has been associated with the destruction of enzyme activity. Blanching is usually carried out between 75 and 95 °C for 1 to 10 minutes, depending on the size of individual vegetable pieces. Blanched vegetables should be promptly cooled down to control and minimize the degradation of soluble and heat-labile nutrients.

The enzymes used as indicators of effectiveness of the blanching treatment are peroxidase, catalase, and more recently lipoxygenase. Peroxidase inactivation is commonly used in vegetable processing, since peroxidase is easily detected and is the most heat stable of these enzymes.

Vegetables can be blanched in hot water, steam, and in the microwave. Hot water blanching is the most common way of processing vegetables. Blanching times recommended for various vegetables are given in Table 6, which indicates that the operation time can vary depending on the intended product use. For water blanching, vegetables are put in a basket and then placed in a kettle of boiling water covered with a lid. Timing begins immediately (Archuleta, 2003). Steam blanching takes longer than the water method, but helps retain water-soluble nutrients such as water-soluble vitamins. For steam blanching, a single layer of vegetables is placed on a rack or in a basket at 3-5 cm above water boiling in a kettle. A tightly fitted lid is placed on the kettle and timing is started. Microwave blanching is usually recommended for small quantities of vegetables prior to freezing. Due to the non-uniform heating disadvantage of microwaves, research is still being conducted to obtain better results with microwave blanching.

Table: Vegetable freezing guide.

Vegetable	Preparation	Blanch/Freeze	
Asparagus	Wash and sort by size. Snap off tough ends. Cut stalks into 5-cm lengths.	Water blanch	2 min
		Steam blanch	3 min
Beans	Wash and trim the ends. Cut if desired.	Water blanch	Steam blanch:
		Whole: 3 min.	Whole: 4 min.
		Cut: 2min.	Cut: 3min.

Beets	Wash and remove the tops leaving 2.5 cm of stem and root.	Cook until tender: 25-30 min Cool promptly, peel, trim. Cut into slices or cubes and pack.	
Broccoli	Wash and cut into pieces.	Water blanch	3 min.
		Steam blanch	3 min.
Cabbage	Wash and cut into wedges.	Water blanch	3 min.
		Steam blanch	4 min.
Carrots	Wash, peel and trim. Cut if desired.	Water blanch 5 min.	
Cauliflower	Discard leaves; steam and wash. Break into flowerets.	Water blanch Whole: 5 min.	Steam blanch Whole: 7 min
Corn	Remove husks and silks. Trim ends and wash.	Water blanch Whole: 5 min.	Steam blanch Whole: 7 min
Greens	Select young tender greens. Wash and trim the leaves.	Water blanch Steam blanch	2 min. 3 min.
Herbs	Wash.	No heat treatment is needed.	
Mushrooms	Wipe and damp with paper towel. Trim hard tip of stems. Sort and cut large mushrooms.	May be frozen without heat treatment.	
Peas	Shell garden peas.	Water blanch 1-1/2 min.	Steam blanch 1-1/2 min.
Peppers	Wash, remove stems and seeds.	Freeze whole or cut as desired. No heat treatment is needed.	
Potatoes	Peel, cut or grate as desired.	Water blanch Whole: 5 min. Pieces: 2-3 min.	

Packaging

There are several factors to consider in packaging frozen vegetables, which include protection from atmospheric oxygen, prevention of moisture loss, retention of flavour, and rate of heat transfer through the package. There are two basic packing methods recommended for frozen vegetables: dry pack and tray pack.

In the dry pack method, the blanched and drained vegetables are put into meal-sized freezer bags and packed tightly to cut down on the amount of air in the package. Proper headspace (approximately 2 cm) is left at the top of rigid containers before closing. For freezer bags, the headspace is larger. Provision for headspace is not necessary for foods such as broccoli, asparagus, and Brussels sprouts, as they do not pack tightly in containers.

In the tray pack method, chilled, well-drained vegetables are placed in a single layer on shallow trays or pans. Trays are placed in a freezer until the vegetables become firm, and then removed. Vegetables are filled into containers. Tray-packed foods do not freeze in a block but remain loosely distributed so that the amount needed can be poured from the container and the package reclosed.

Drying of Fruits and Vegetables

Food drying is one of the oldest methods of preserving food for later use. It can either be an alternative to canning or freezing, or compliment these methods. Drying foods is simple, safe and easy to learn.

How Drying Preserves Food

Drying removes the moisture from the food so bacteria, yeast and mold cannot grow and spoil the food. Drying also slows down the action of enzymes (naturally occurring substances which cause foods to ripen), but does not inactivate them.

Because drying removes moisture, the food becomes smaller and lighter in weight. When the food is ready for use, the water is added back, and the food returns to its original shape. Foods can be dried in the sun, in an oven or in a food dehydrator by using the right combination of warm temperatures, low humidity and air current.

In drying, warm temperatures cause the moisture to evaporate. Low humidity allows moisture to move quickly from the food to the air. Air current speeds up drying by moving the surrounding moist air away from the food.

Drying Foods Out-of-Doors Sun Drying

The high sugar and acid content of fruits make them safe to dry in the sun. Vegetables and meats are not recommended for sun drying. Vegetables are low in sugar and acid. This increases the risks for food spoilage. Meats are high in protein making them ideal for microbial growth when heat and humidity cannot be controlled.

To dry in the sun, hot, dry, breezy days are best. A minimum temperature of 86°F is needed with higher temperatures being better. It takes several days to dry foods out-of-doors. Because the weather is uncontrollable, sun drying can be risky. Also, the high humidity in the South is a problem. Humidity below 60 percent is best for sun drying. Often these ideal conditions are not available when fruit ripens.

Fruits dried in the sun are placed on trays made of screen or wooden dowels. Screens need to be safe for contact with food. The best screens are stainless steel, teflon coated fiberglass or plastic. Avoid screens made from "hardware cloth." This is galvanized metal cloth that is coated with cadmium or zinc. These materials can oxidize, leaving harmful residues on the food. Also avoid copper and aluminium screening. Copper destroys vitamin C and increases oxidation. Aluminium tends to discolor and corrode.

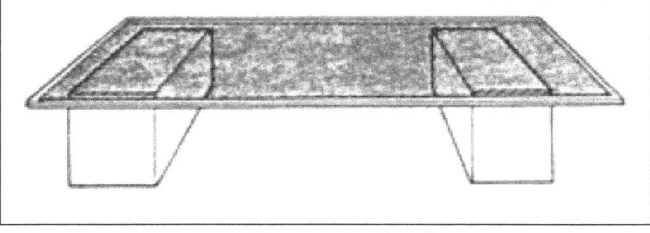

Outdoor Drying Rack

Most woods are fine for making trays. However, do not use green wood, pine, cedar, oak or red-wood. These woods warp, stain the food or cause off-flavors in the food. Place trays on blocks to allow for better air movement around the food. Because the ground may be moist, it is best to place the racks or screens on a concrete driveway or if possible over a sheet of aluminum or tin. The reflection of the sun on the metal increases the drying temperature. Cover the trays with cheesecloth to help protect the fruit from birds or insects. Fruits dried in the sun must be covered or brought under shelter at night. The cool night air condenses and could add moisture back to the food, thus slowing down the drying process.

Solar Drying

Recent efforts to improve on sun drying have led to solar drying. Solar drying also uses the sun as the heat source. A foil surface inside the dehydrator helps to increase the temperature. Ventilation speeds up the drying time. Shorter drying times reduce the risks of food spoilage or mold growth.

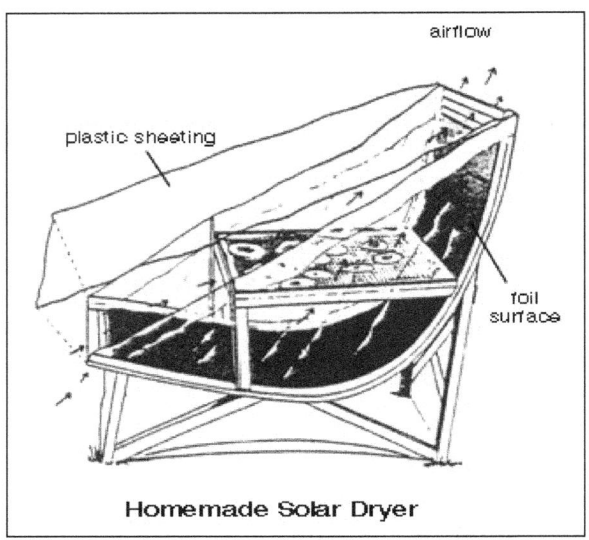

Homemade Solar Dryer

Pasteurization

Sun or solar dried fruits and vine dried beans need treatment to kill any insect and their eggs that might be on the food. Unless destroyed, the insects will eat the dried food. There are two recommended pasteurization methods:

- Freezer Method: Seal the food in freezer-type plastic bags. Place the bags in a freezer set at 0 °F or below and leave them at least 48 hours.

- Oven Method: Place the food in a single layer on a tray or in a shallow pan. Place in an oven preheated to 160 °F for 30 minutes.

After either of these treatments the dried fruit is ready to be conditioned and stored.

Drying Foods Indoors

Most foods can be dried indoors using modern dehydrators, convection ovens or conventional ovens. Microwave ovens are recommended only for drying herbs, because there is no way to create airflow in them.

Food Dehydrators

Food dehydrator is a small electrical appliance for drying food indoors. A food dehydrator has an electric element for heat and a fan and vents for air circulation. Dehydrators are efficiently designed to dry foods quickly at 140 °F.

Food dehydrators are a relatively new item and are available from department stores, mail-order catalogs, natural food stores, seed catalogs and garden supply stores. Costs vary from $40 to $350 or above depending on features. Some models are expandable and additional trays can be purchased later. Twelve square feet of drying space dries about a half-bushel of produce.

Oven Drying

Everyone who has an oven has a dehydrator. By combining the factors of heat, low humidity and air flow, an oven can be used as a dehydrator. An oven is ideal for occasional drying of meat jerkies, fruit leathers, banana chips or for preserving excess produce like celery or mushrooms. Because the oven is needed for every day cooking, it may not be satisfactory for preserving abundant garden produce.

Oven drying is slower than dehydrators because it does not have a built-in fan for the air movement. (However, some convection ovens do have a fan). It takes about two times longer to dry food in an oven than it does in a dehydrator. Thus, the oven is not as efficient as a dehydrator and uses more energy.

To Use Your Oven - First, check the dial and see if it can register as low as 140 °F. If your oven does not go this low, then your food will cook instead of dry. Use a thermometer to check the temperature at the "warm" setting.

For air circulation, leave the oven door propped open two to six inches. Circulation can be improved by placing a fan outside the oven near the door. Because the door is left open, the temperature will vary. An oven thermometer placed near the food gives an accurate reading. Adjust the temperature dial to achieve the needed 140 °F.

Drying trays should be narrow enough to clear the sides of the oven and should be 3 to 4 inches shorter than the oven from front to back. Cake cooling racks placed on top of cookie sheets work well for some foods. The oven racks, holding the trays, should be two to three inches apart for air circulation.

Drying Fruits

Dried fruits are unique, tasty and nutritious. Begin by washing the fruit and coring it, if needed. For drying, fruits can be cut in half or sliced. Some can be left whole. Thin, uniform, peeled slices dry the fastest. The peel can be left on the fruit, but unpeeled fruit takes the longer to dry. Apples can be cored and sliced in rings, wedges, or chips. Bananas can be sliced in coins, chips or sticks.

Fruits dried whole take the longest to dry. Before drying, skins need to be "checked" or cracked to speed drying. To "check" the fruit place it in boiling water and then in cold water. Because fruits contain sugar and are sticky, spray the drying trays with non-stick cooking spray before placing the fruit on the trays. After the fruit dries for one to two hours, lift each piece gently with a spatula and turn.

Pre-treating the Fruit

Pre-treatments prevent fruits from darkening. Many light-colored fruits, such as apples, darken rapidly when cut and exposed to air. If not pre-treated, these fruits will continue to darken after they have dried.

For long-term storage of dried fruit, sulfuring or using a sulfite dip are the best pretreatments. However, sulfites found in the food after either of these treatments have been found to cause asthmatic reactions in a small portion of the asthmatic population. Thus, some people may want to use the alternative shorter-term pretreatments. If home dried foods are eaten within a short time, there may be little difference in the long- and short-term pretreatments.

- Sulfuring: Sulfuring is an old method of pre-treating fruits Sublimed sulfur is ignited and burned in an enclosed box with the fruit. The sulfur fumes penetrate the fruit and act as a pre-treatment by retarding spoilage and darkening of the fruit. Fruits must be sulfured out-of-doors where there is adequate air circulation.

- Sulfite Dip: Sulfite dips can achieve the same long-term anti-darkening effect as sulfuring, but more quickly and easily. Either sodium bisulfite, sodium sulfite or sodium meta-bisulfite that are USP (food grade) or Reagant grade (pure) can be used. To locate these, check with your local drugstores or hobby shops, where wine-making supplies are sold.

  ○ Directions for Use: Dissolve 3⁄4 to 1 1⁄2 teaspoons sodium bisulfite per quart of water. (If using sodium sulfite, use 1 1⁄2 to 3 teaspoons. If using sodium metabisulfite, use 1 to 2 tablespoons.) Place the prepared fruit in the mixture and soak 5 minutes for slices, 15 minutes for halves. Remove fruit, rinse lightly under cold water and place on drying trays. Sulfited foods can be dried indoors or outdoors.

- Ascorbic Acid: Ascorbic acid (vitamin C) mixed with water is a safe way to prevent fruit browning. However, its protection does not last as long as sulfuring or sulfiting. Ascorbic acid is available in the powdered or tablet form, from drugstores or grocery stores. One teaspoon of powdered ascorbic acid is equal to 3000 mg of ascorbic acid in tablet form. (If you buy 500 mg tablets, this would be six tablets).

  ○ Directions for Use: Mix 1 teaspoon of powdered ascorbic acid (or 3000 mg of ascorbic acid tablets, crushed) in 2 cups water. Place the fruit in the solution for 3 to 5 minutes. Remove fruit, drain well and place on dryer trays. After this solution is used twice, add more acid.

  ○ Ascorbic Acid Mixtures: Ascorbic acid mixtures are a mixture of ascorbic acid and sugar sold for use on fresh fruits and in canning or freezing. It is more expensive than and not as effective as using pure ascorbic acid.

- ◦ Directions for Use: Mix 1 1/2 tablespoons of ascorbic acid mixture with one quart of water. Place the fruit in the mixture and soak 3 to 5 minutes. Drain the fruit well and place on dryer trays. After this solution is used twice, add more ascorbic acid mixture.

- Fruit Juice Dip: A fruit juice that is high in vitamin C can also be used as a pretreatment, though it is not as effective as pure ascorbic acid. Juices high in vitamin C include orange, lemon, pineapple, grape and cranberry. Each juice adds its own color and flavor to the fruit.

  - ◦ Directions for Use: Place enough juice to cover fruit in a bowl. Add cut fruit. Soak 3 to 5 minutes, remove fruit, drain well and place on dryer trays. This solution may be used twice, before being replaced. (The used juice can be consumed.)

- Honey Dip: Many store-bought dried fruits have been dipped in a honey solution. A similar dip can be made at home. Honey dipped fruit is much higher in calories.

  - ◦ Directions for Use: Mix 1/2 cup sugar with 1 1/2 cups boiling water. Cool to lukewarm and add 1/2 cup honey. Place fruit in dip and soak 3 to 5 minutes. Remove, drain well and place on dryer trays.

- Syrup Blanching: Blanching fruit in syrup helps it retain color fairly well during drying and storage. The resulting product is similar to candied fruit. Fruits that can be syrup blanched include apples, apricots, figs, nectarines, peaches, pears, plums and prunes.

  - ◦ Directions for Use: Combine 1 cup sugar, 1 cup light corn syrup and 2 cups water in a saucepot. Bring to a boil. Add 1 pound of prepared fruit and simmer 10 minutes. Remove heat and let fruit stand in hot syrup for 30 minutes. Lift fruit out of syrup, rinse lightly in cold water, drain on paper toweling and place on dryer trays.

- Steam Blanching: Steam blanching also helps retain color and slow oxidation. However, the flavor and texture of the fruit is changed.

  - ◦ Directions: Place several inches of water in a large saucepot with a tight fitting lid. Heat to boiling. Place fruit not more than 2 inches deep, in a steamer pan or wire basket over boiling water. Cover tightly with lid and begin timing immediately. See below for blanching times. Check for even blanching half way through the blanching time. Some fruit may need to be stirred. When done, remove excess moisture using paper towels and place on dryer trays.

Drying the Prepared Fruit

Whichever drying method you choose-sun drying, solar drying, oven drying or dehydrator drying-be sure to place the fruit in a single layer on the drying trays. The pieces should not touch or overlap. Follow the directions for the drying method you choose and dry until the food tests dry. Approximate drying times are given below. Food dries much faster at the end of the drying period, so watch it closely.

Determining Dryness of Fruits

Since dried fruits are generally eaten without being rehydrated, they should not be dehydrated to the point of brittleness. Most fruits should have about 20 percent moisture content when dried. To test for dryness, cut several cooled pieces in half. There should be no visible moisture and you should not be able to squeeze any moisture from the fruit. Some fruits may remain pliable, but are not sticky or tacky. If a piece is folded in half, it should not stick to itself. Berries should be dried until they rattle when shaken.

After drying, cool fruit 30 to 60 minutes before packaging. Packaging food warm can lead to sweating and moisture buildup. However, excessive delays in packaging could allow moisture to re-enter food. Remember, if you have dried fruit in the sun, it must be pasteurized before it is packaged.

Conditioning Fruits

When dried fruit is taken from the dehydrator or oven, the remaining moisture may not be distributed equally among the pieces because of their size or their location in the dehydrator. Conditioning is a process used to equalize the moisture and reduce the risk of mold growth.

To condition the fruit, take the dried fruit that has cooled and pack it loosely in plastic or glass jars. Seal the containers and let them stand for seven to ten days. The excess moisture in some pieces will be absorbed by the drier pieces. Shake the jars daily to separate the pieces and check the moisture condensation. If condensation develops in the jar, return the fruit to the dehydrator for more drying. After conditioning, package and store the fruit.

Drying Vegetables

Vegetables can also be preserved by drying. Because they contain less acid than fruits, vegetables are dried until they are brittle. At this stage, only 10% moisture remains and no microorganism can grow.

Preparing Vegetables

To prepare vegetables for drying, wash in cool water to remove soil and chemical residues. Trim, peel, cut, slice or shred vegetables according to the directions for each vegetable in the chart below. Remove any fibrous or woody portions and core when necessary, removing all decayed and bruised areas. Keep pieces uniform in size so they will dry at the same rate. A food slicer or food processor can be used. Prepare only as many as can be dried at one time.

Pre-treating Vegetables

Blanching is a necessary step in preparing vegetables for drying. By definition, blanching is the process of heating vegetables to a temperature high enough to destroy enzymes present in tissue. Blanching stops the enzyme action which could cause loss of color and flavor during drying and storage. It also shortens the drying and rehydration time by relaxing the tissue walls so moisture can escape and later re-enter more rapidly.

Vegetables can be water blanched or steam blanched. Water blanching usually results in a greater loss of nutrients, but it takes less time than steam blanching.

- Water Blanching - Fill a large pot 2/3 full of water, cover and bring to a rolling boil. Place the vegetables in a wire basket or a colander and submerge them in the water. Cover and blanch according to directions. Begin timing when water returns to boiling. If it takes longer than one minute for the water to come back to boiling, too many vegetables were added. Reduce the amount in the next batch.

- Steam Blanching - Use a deep pot with a tight fitting lid and a wire basket, colander or sieve placed so the steam will circulate freely around the vegetables. Add water to the pot and bring to a rolling boil. Place the vegetables loosely in the basket no more than 2 inches deep. Place the basket of vegetables in the pot, making sure the water does not come in contact with the vegetables. Cover and steam according to the directions.

Cooling and Drying the Prepared Vegetables

After blanching, dip the vegetables briefly in cold water. When they feel only slightly hot to the touch, drain the vegetables by pouring them directly onto the drying tray held over the sink. Wipe the excess water from underneath the tray and arrange the vegetables in a single layer. Then place the tray immediately in the dehydrator or oven. The heat left in the vegetables from blanching will cause the drying process to begin more quickly. Watch the vegetables closely at the end of the drying period. They dry much more quickly at the end and could scorch.

Determining Dryness of Vegetables

Vegetables should be dried until they are brittle or "crisp." Some vegetables would actually shatter if hit with a hammer. At this stage, they should contain about 10 percent moisture. Because they are so dry, they do not need conditioning like fruits.

Drying Fruit Leather

Fruit leather is a tasty, chewy, dried fruit product. Fruit leathers are made by pouring puréed fruit onto a flat surface for drying. When dried, the fruit is pulled from the surface and rolled. It gets the name "leather" from the fact that when puréed fruit is dried, it is shiny and has the texture of leather.

Leather from Fresh Fruit

- Select ripe or slightly overripe fruit.

- Wash fresh fruit or berries in cool water. Remove peel, seeds and stem.

- Cut fruit into chunks. Use 2 cups of fruit for each 13" x 15" inch fruit leather. Purée fruit until smooth.

- Add 2 teaspoons of lemon juice or 1/8 teaspoon ascorbic acid (375 mg) for each 2 cups of light colored fruit to prevent darkening.

- To sweeten, add corn syrup, honey or sugar. Corn syrup or honey is best for longer storage because it prevents crystals. Sugar is fine for immediate use or short storage. Use 1/4 to 1/2 cup sugar, corn syrup or honey for each 2 cups of fruit. Saccharin-based sweeteners could also be used to reduce tartness without adding calories. Aspartame sweeteners may lose sweetness during drying.

Leathers from Canned or Frozen Fruits

- Home preserved or store bought canned or frozen fruit can be used.

- Drain fruit, save liquid.

- Use 1 pint of fruit for each 13" x 15" leather.

- Purée fruit until smooth. If thick, add liquid.

- Add 2 teaspoons of lemon juice or 1/8 teaspoon ascorbic acid (375 mg) for each 2 cups of light colored fruit to prevent darkening.

- If desired, sweeten as directed above for leathers from fresh fruit.

- Applesauce can be dried alone or added to any fresh fruit purée as an extender. It decreases tartness and makes leather smoother and more pliable.

Drying the Leather

For drying in the oven or sun, line cookie sheets with plastic wrap. In a dehydrator, use plastic wrap or the specially designed plastic sheets that come with the dehydrator. Pour the leather onto the lined cookie sheets or tray. Spread it evenly to a thickness of 1/8 inch.

Dry the fruit leather at 140° F until no intention is left when you touch the center with your finger. This could take about 6 to 8 hours in the dehydrator, up to 18 hours in the oven and 1 to 2 days in the sun. While still warm, peel from the plastic wrap. Cool and rewrap in plastic and store.

Packaging and Storing Dried Foods

After foods are dried, cool them completely. Then package them in clean moisture-vapor-resistant containers. Glass jars, metal cans or freezer containers are good storage containers, if they have tight-fitting lids. Plastic freezer bags are acceptable, but they are not insect and rodent proof. Fruit that has been sulfured or sulfited should not touch metal. Place the fruit in a plastic bag before storing it in a metal can.

Dried food should be stored in a cool, dry, dark place. Most dried fruits can be stored for 1 year at 60° F, 6 months at 80 °F. Dried vegetables have about half the shelf-life of fruits. Fruit leathers should keep for up to 1 month at room temperature. To store any dried product longer, place it in the freezer.

Using Dried Foods

Dried fruits can be eaten as is or reconstituted. Dried vegetables must be reconstituted. Once reconstituted, dried fruits or vegetables are treated as fresh. Fruit leathers and meat jerky are eaten as is. To reconstitute dried fruits or vegetables, add water to the fruit or vegetable and soak until the desired volume is restored. (See the chart on rehydrating dried food, for the amount of water to add and minimum soaking time.) Do not over-soak the food. Over-soaking produces loss of flavor and a mushy, water-logged texture.

For soups and stews, add the dehydrated vegetables, without rehydrating them. They will rehydrate as the soup or stew cooks. Also, leafy vegetables and tomatoes do not need soaking. Add enough water to cover and simmer until tender. If soaking takes more than 2 hours, refrigerate the product for the remainder of the time.

Canning of Fruits and Vegetables

Canning is a method of preserving food in which the food contents are processed and sealed in an airtight container. Canning provides a shelf life typically ranging from one to five years, although under specific circumstances it can be much longer. It is also known as Appertisation. This practice removes oxygen, destroys enzymes, prevents the growth of undesirable bacteria, yeasts, moulds and helps forming high vacuum in cans.

Canned products.

Principles of Canning

Destruction of spoilage organisms within the sealed container by means of heat.

- Many vegetables begin losing some of their vitamins when harvested. Nearly half the vitamins may be lost within a few days unless the fresh produce is cooled or preserved. Within 1 to 2 weeks, even refrigerated produce loses half or more of some of its vitamins.

- The heating process during canning destroys from one-third to one-half of vitamins A and C, thiamin, and riboflavin.

Canning Process

Flow chart of canning process for fruits and vegetables.

Types of Canning

Canning is two types: Water Bath Canning and Pressure Canning.

Water Bath Canning

- Used for high acid food (pH 4.6 and below).

- Uses boiling water temperature of 100 °C.

- Shorter times, usually from 5 – 85 minutes. E.g. generally for all fruits.

Water bath canning.

Pressure Canning

- Use for low acid food. Uses high temperature ranging from 116 °C - 121 °C.

- Longer times, usually from 20 - 100 minutes e.g. generally all vegetables, meats, poultry, soups etc.

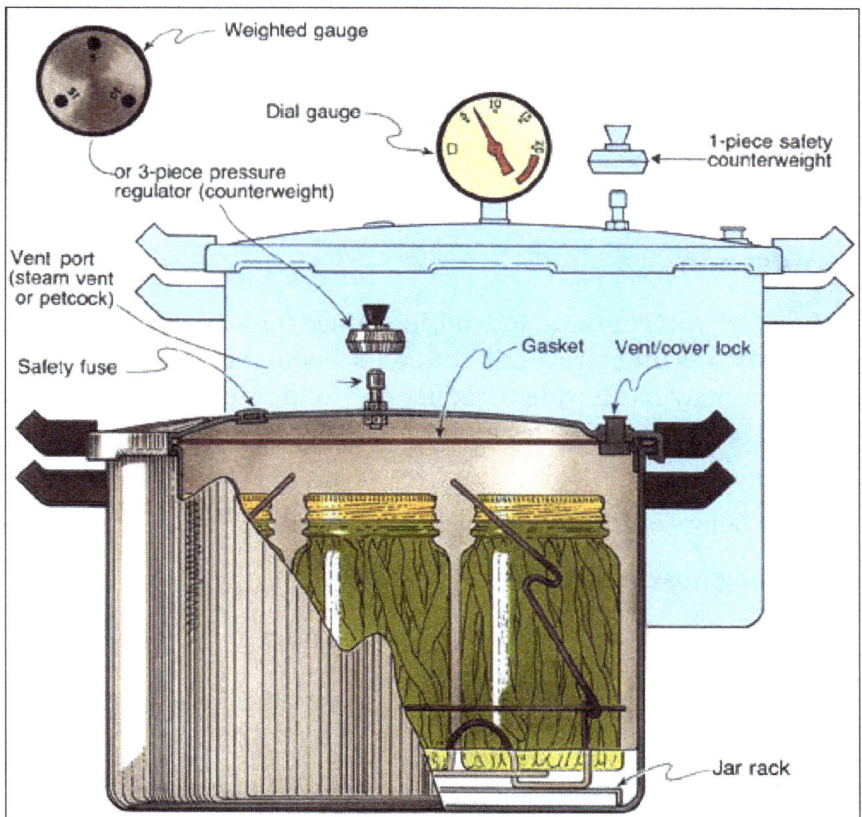

Pressure canning.

Advantages of Canning

- Helps save money.

- Saves nutritional value.

- Can last for years.

- Does not require electricity or refrigeration to store.

- Preservative- and pesticides-free food.

- Offseason availability.

Disadvantages of Canning

- Time-consuming process.

- Initial startup cost of buying the equipments.

- Person should know which foods are high risk and which are low.

- Change in ingredient may require a change in processing.

- Canning machines are expensive.

Market Trends

Canned/preserved fruit, vegetables, beans and other canned/preserved food have small but significant sales within the category whereas sales of canned/preserved fish/seafood, meat and meat products, and tomatoes are low or negligible, as consumers prefers fresh produce, which is very readily available.

Competitive Landscape

The Oudh Sugar Mills Ltd will continue to lead in canned/preserved food with a value share of 10% in 2014, followed by Tai Industries Ltd and MTR Foods Ltd, with value shares of 10% and 9% respectively. The company has a wider product portfolio catering to affluents with its Morton brand. Morton has strong retail value sales among all formats of retail chains and also has significant institutional sales.

Factors to keep in mind before starting canning centre:

- Start Your Canning Business Small.

- Develop Your Canning place.

- Access Quality Canning Ingredients.

- Price Canned Food Accurately.

- Sell Your Canning Story.

- Manage Your Time and Resources.

- Market Your Canning Business Year-round.

Storage of Fruits and Vegetables

Food Produce does not improve in storage. Basic aim of storage is to slow down the ageing process caused due to respiration, moisture loss and disease decay.

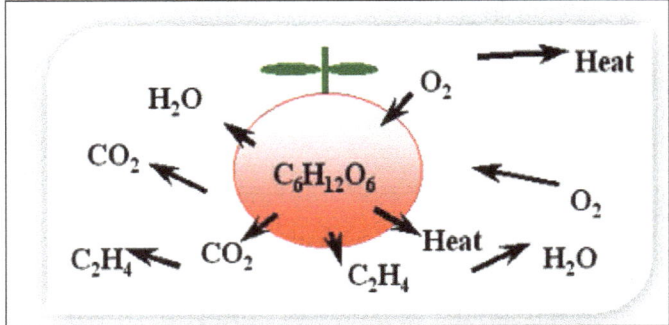

Mechanism of ageing and food loss.

All fruit and vegetables continue to breathe (respire) after harvest. Respiration is the chemical process by which fruits and vegetables convert sugars and oxygen into carbon dioxide, water, and heat [carbohydrates (sugars) + oxygen = carbon dioxide + water + heat]. Fruits and vegetables are classified on the basis of their rate of respiration.

Table: Classification of horticultural commodities according to their relative rates of respiration.

Class	Respiration range (mg CO_2 Kg^{-1} h^{-1})	Class
Very low	<5	Dates, nuts, dried fruits and vegetable.
Low	5-10	Apple, citrus fruits, grape, kiwifruit, honey-dew melon, water melon, papaya, persimmon, pineapple, beat, celery, cranberry, garlic, onion, potato (mature), sweet potato.
Moderate	10-20	Apricot, banana, cherry, blueberry, nectarine, pear, plum, fig, gooseberry, mango, peach, summer squash, cantaloupe, celeriac, cucumber, lettuce (head), olive, potato, (immature), radish (topped), tomato, cabbage, pepper.
High	20-40	Avocado, blackberry, carrot (with tops), strawberry, raspberry.
Very high	40-60	Artichoke, bean sprouts, brussels, sprouts, snap bean, watercress, endive, green onions, kale, okra.
Extremely High	>60	Asparagus, broccoli, mushroom, parsley, peas, spinach, sweet corn.

Respiration rate depends on temperature, oxygen (O_2), carbon Dioxide (CO_2), humidity (RH), mechanical Injury (cuts & bruises) and disease and pest infections. The relative perishability and storage life of certain fresh horticultural crops are given in table.

Role of Temperature in Food Storage

Temperature is the most important environmental factor because it has a profound effect on the rates of biological reactions. Within the physiological range of temperatures (0 -30 °C), the velocity of a biological reaction increases 2 to 3 fold per every 10 °C rise in temperature (Q_{10}). The Q_{10} can be calculated by dividing the reaction rate at a higher temperature by the rate at a 10 °C low temperature.

$$Q_{10} \frac{\text{Rate of deterioration at temperature T} + 10°C \text{ (R2)}}{\text{Rate of deterioration at temperature T (R1)}}$$

Table: Classification of fresh horticultural crops according to their relative perishability and potential storage life in air at near optimum temperature and relative humidity.

Relative perishability	Potential storage life (weeks)	Commodities
Very high	<2	Apricot, blackberry, blueberry, cherry, fig, raspberry, strawberry; asparagus, bean sprouts, broccoli, cauliflower, green onion, leaf lettuce, mushroom, muskmelon, pea, spinach, sweet corn, tomato (ripe); most cut flowers and foliage; minimally processed fruits and vegetables.

High	2-4	Avocado, banana, grape (without SO_2 treatment), guava, loquat, mandarin, mango, melons (honeydew, crenshaw, Persian), nectarine, papaya, peach, plum; artichoke, green beans, Brussels sprouts, cabbage, celery, eggplant, head lettuce, okra, pepper, summer squash, tomato (partially ripe).
Moderate	4- 8	Apple and pear (some cultivars), grape (SO_2-treated), orange, grape-fruit, lime, kiwifruit, persimmon, pomegranate; table beet, carrot, radish, potato (immature).
Low	8-16	Apple and pear (some cultivars), lemon; potato (mature), dry onion, garlic, pumpkin, winter squash, sweet potato, taro, yam; bulbs and other propagules of ornamental plants.
Very low	>16	Tree nuts, dried fruits and vegetables.

The Q_{10} allows us to calculate the expected respiration rates at one temperature from a known rate at another temperature. However, Q_{10} for respiration vary with smaller at higher temperatures and greater at lower temperatures.

Effect of Temperature on the Deterioration Rate of a Non-chilling Sensitive Commodity

Using Q_{10} values the effect of different temperatures on the rates of respiration or deterioration and relative shelf life of a typical perishable commodity can be estimated. For example if a commodity has a mean shelf life of 13 days at 20 °C then it can be stored for as long as 100 days at °C, but will last no more than 4 days at 40 °C.

Table: Effect of increasing temperature on Q_{10} values of respiration.

Temperatures (°C)	Q10 Values)
0-10	2.5 to 4.0
10-20	2.0 to 2.5
20-30	1.5 to 2.0
30-40	1.0 to 1.5

Table: Effect of temperature on the deterioration rate of a non-chilling sensitive commodity.

Temperature	Assumed Q_{10}	Relative Velocity of deterioration	Relative postharvest life	Loss per day
0		1.0	100	1
10	3.0	3.0	33	3
20	2.5	7.5	13	8
30	2.0	15.0	7	4
40	1.5	22.5	5	25

Recommended storage temperatures for a selection of fruits and vegetables may vary according to their chemical composition rate of reparation. Storage of fruits and vegetables below their optimum temperature may cause physiological injury and shorten their storage life.

Table: Storage Requirements for Common Vegetables and Fruit.

Produce	Temperature (°C)	Relative Humidity	Storage Life
Apples	32	90 − 95%	4 − 6 months
Beets	32	90%	1 − 3 months
Bressels Sprouts	32	90 − 95%	3 − 5 weeks
Cabbage	32	90 − 95%	3 − 4 months
Carrot	32	90 − 95%	4 − 6 months
Cauliflower	32	90 − 95%	2 − 4 weeks
Celeriac	32	90 − 95%	3 − 4 months
Chinese Cabbage	32	90 − 95%	1 − 2 months
Dry beans	32 − 50	65 − 70%	1 year
Garlic	32	65 − 70%	6 − 7 months
Horseradish	30 − 32	90 − 95%	10 − 12 months
Kale	32	90 − 95%	10 − 14 days
Kohlrabi	32	90 − 95%	2 − 4 weeks
Leeks	32	90 − 95%	1 − 3 months
Onios	32	65 − 70%	5 − 8 months
Parsnips	32	90 − 95%	2 − 6 months
Pears	32	90 − 95%	1 − 2 months
Sweet Pepper	45 − 50	90 − 95%	8 − 10 days
Potatoes	38 − 40	90%	5 − 8 months
Pumpkins	50 − 55	70 − 75%	2 − 3 months
Rutabaga	32	90 − 95%	2 − 4 months
Salsify	32	90 − 95%	2 − 4 months
Sweet Potato	55 − 60	85 − 90%	4 − 6 months
Tomatoes (Green)	55 − 60	85 − 90%	2 − 6 weeks
Turnips	32	90 − 95%	4 − 5 months

Table: Susceptibility of fruits and vegetables to chilling injury at low but non-freezing temperatures.

Commodity	Lowest safe temperature (°C)	Chilling injury symptoms
Aubergines	7	Surface scald, *Alternaria* rot
Avocados	5-13	Grey discoloration of flesh
Bananas (green/ ripe)	12-14	Dull, grey-brown skin color
Beans (green)	7	Pitting, russeting
Cucumbers	7	Pitting water-soaked spots, decay
Grapefruit	10	Brown scald, piking, watery breakdown
Lemons	13-15	Pitting, membrane stain, red blotch

Limes	7-10	Pitting
Mangoes	10-13	Grey skin scald, uneven ripening
Melons: Honeydew	7-10	Pitting failure to ripen, decay
Watermelon	5	Pitting, biker flavour
Okra	7	Discoloration, water-soaked areas, piking
Oranges	7	Pitting brown stain, watery breakdown
Papaya	7	Pitting failure to ripen, off-flavour, decay
Pineapples	7-10	Dull green color, poor flavour
Potatoes	4	Internal discoloration, sweetening
Pumpkins	10	Decay
Sweet peppers	7	Pitting, *Alternaria* rot
Sweet potato	13	Internal discoloration, piking, decay
Tomatoes: Mature green	13	Water-soaked softening, decay
Ripe	7-10	Poor color, abnormal ripening, *Alternaria* rot

Moisture Loss and Food Storability

Water is the main constituent of fresh fruits and vegetables which maintains the freshness of the produce by maintaining the turgidity of the cells. Harvested produce should be handled carefully to minimize both water loss and the presence of free water. Water loss results in weight loss, wilting and shriveling, while free water or condensation facilitates pathogen growth. Understanding and managing water relations is therefore a critical component of postharvest handling, second only to temperature management. At the time of harvest the water content of fruits and vegetables is very high and produce has a fresh appearance and crisp texture. Harvesting removes the plant part from its water supply and the product begins to lose weight. This loss of water has an immediate economic effect in that it reduces saleable weight. Continued water loss causes wilting and/or shriveling. Consequently, reducing water loss improves produce appearance, quality, shelf life and profitably. During postharvest handling and storage, fresh fruits and vegetables lose moisture through their skins via the transpiration process. This process includes:

- The transport of moisture through the skin of the commodity.

- The evaporation of this moisture from the commodity surface.

- The convective mass transport of the moisture to the surroundings.

Moisture loss from a fruit or vegetable is driven by a difference in water vapor pressure between the product surface and the environment. Evaporation which occurs at the product surface is an endothermic process which will cool the surface, thus lowering the vapor pressure at the surface and reducing transpiration. Respiration within the fruit or vegetable, on the other hand, tends to increase the product's temperature, thus raising the vapor pressure at the surface and increasing transpiration. In addition, factors such as surface structure, skin permeability, and air flow also affect the transpiration rate. Thus, it can be seen that within fruits and vegetables, complex heat and mass transfer phenomena occur, which must be considered when evaluating the transpiration rates of commodities.

Table: Water content (%) by weight of some common fruits and vegetables.

Fruit	Water content (%)	Vegetable	Water content (%)
Apple	84	Asparagus	93
Avocado	76	Beans (green)	89
Banana (green)	76	Broccoli	90
Blue berry	83	Brussels sprouts	85
Cantaloupe	93	Cabbage	92
Cherry	80	Carrot	88
Citrus	89	Cauliflower	92
Grape	82	Lettuce	95
Grape fruit	89	Mushroom	91
Honeydew melon	93	Onion (dry)	88
Kiwifruit	82	Pepper (sweet)	92
Mango	82	Potato	78
Orange	86	Pumpkin	91
Peach	89	Spinach	93
Pear	83	Squash (summer)	94
Plum	87	Squash (winter)	85
Watermelon	93	Tomato (firm ripe)	94

Controlled Atmosphere

Quality and the freshness of fruit and vegetables are retained under Controlled Atmosphere conditions without the use of any chemicals. Under CA conditions, many products can be stored for 2 to 4 times longer than usual. Controlled atmosphere storage is a system for holding produce in an atmosphere that differs substantially from normal air in respect to CO_2 and O_2 levels. Controlled atmosphere storage refers to the constant monitoring and adjustment of the CO_2 and O_2 levels within gas tight stores or containers. The gas mixture will constantly change due to metabolic activity of the respiring fruits and vegetables in the store and leakage of gases through doors and walls. The gases are therefore measured periodically and adjusted to the predetermined level by the introduction of fresh air or nitrogen or passing the store atmosphere through a chemical to remove CO_2. There are different types of controlled atmosphere storage depending mainly on the method or degree of control of the gases. Some researchers prefer to use the terms"static controlled atmosphere storage" and "flushed controlled atmosphere storage" to define the two most commonly used systems. "Static" is where the product generates the atmosphere and "flushed" is where the atmosphere is supplied from a flowing gas stream, which purges the store continuously. Systems may be designed which utilize flushing initially to reduce the O_2 content then either injecting CO_2 or allowing it to build up through respiration, and then maintenance of this atmosphere by ventilation and scrubbing.

Table: Maximum Permissible limit of Water Loss in certain fruits and vegetables.

S. No.	Commodity	Maximum Permissible water loss	S. No.	Commodity	Maximum Permissible water loss
	Fruits		17	Cabbage	6-11
1	Apple	7.5	18	Carrot	8
2	Blackberry	6	19	Carrot with leaves	4
3	Grape	5	20	Cauliflower	7
4	Nectarine	21	21	Celery	10
5	Papaya	7	22	Cucumber	5
6	Peach	11-16.4	23	Leaf lettuce	3-5
7	Pear	6	24	Leek	7
8	Persimmon	7-13	25	Onion	10
9	Raspberry	6	26	Pepper green	8
	Vegetables		27	Potato	7
10	Asparagus	8	28	Spinach	3
11	Beans, Broad	6	29	Squash, summer	24
12	Beans, runner	5	30	Squash, winter	15
13	Beetroot	7	31	Sweet corn	7
14	Beetroot with tops	5	32	Tomato	4-7
15	Broccoli	4	33	Turnip with leaves	5
16	Brussels sprout	8	34	Water cress	7

Table: Recommended conditions for controlled atmosphere storage. Only fruits and vegetables for which commercial uses of C.A. storage are common have been included.

	Temperature range		%O_2 Range	%CO_2 range
	C	F		
Strawberry	0-5	32-41	10	15-20
Apple	0-5	32-41	2-3	1-2
Kiwifruit	0-5	32-41	2	5
Nuts and dried fruits	0-25	32-77	0-1	0-100
Bananas	12-15	54-59	2-5	2-5
Cantaloupe	3-7	38-45	3-5	10-15
Lettuce	0-5	32-41	2-5	0
Mature green	12-20	54-68	3-5	0
Partially-ripe	8-12	47-54	3-5	0

Table: A summary of controlled atmosphere (CA) requirements and recommendations for apples.

Cultivar	O_2 %	CO_2 %	Temp.(°C)	Storage (Month)
Braeburn	1.8	1.0	0.7	6-9

Fuji	1.4	1.0	0.3	7-11
Gala	1.7	1.6	1.3	2-9
Golden Delicious	1.6	2.3	0.5	7-11
Granny Smith	1.4	2.0	0.6	7-11
Idared	2.1	2.5	1.9	7-10
Jonagold	1.4	2.7	0.9	5-10
McIntosh	2.1	2.9	2.5	5-10
Red Delicious	1.6	1.8	0.0	6-11
Royal Gala	1.7	1.8	-0.2	5-8
Average	1.7	2.0	0.9	-

Zero Energy Cool Chambers (ZECC)

Based on the principles of direct evaporative cooling zero energy cool chambers (ZECC) have been developed. The main advantage of this on-farm low cost cooling technology are it does not require any electricity or power to operate and materials required to construct this like bricks, sand bamboo, etc. available easily and cheaply. It is a double brick-wall structure. The cavity is filled with sand and walls of the chamber are soaked in water. Even unskilled labour can build the chamber as it does not require any specialized skill. ZECC can reduce temperature by 10-15 °C and maintain high humidity of about 95% that can increase shelf life and retain quality of horticultural produce. Small and marginal farmers can store a few days harvest to avoid middle men. National Horticulture Board I giving 100% grant in aid for the benefit of the farmers.

Construction of ZECC

- Select an upland having a nearby source of water supply.

- Make floor with brick 165 cm × 115 cm.

- Erect the double wall to a height of 67.5 cm leaving a cavity of 7.5 cm.

- Drench the chamber with water.

- Soak the fine river bed sand with water.

- Fill the 7.5 cm cavity between the double wall with this wet sand.

- Make a frame of top cover with bamboo (165 cm × 115 cm) frame and sirki, straw or dry grass etc.

- Make attach/ shed over the chamber in order to project it from direct sun or rain Operation.

- Keep the sand, bricks and top cover of the chamber wet with water.

- Water twice daily in order to achieve desired temperature and relative humidity or fix a drip system with plastic pipes and micro tubes connected to an overhead water source.

- Store the fruits and vegetables in this chamber by keeping in perforated plastic crates.

- Cover these crates with a thin polyethylene sheet.

Precautions in Construction of ZECC

- Try to site in a place where breezes blow.

- Build in an elevated place to avoid water logging.

- Use clean, unbroken bricks with good porosity.

- Sand should be clean and free of organic matters, clay etc.

- Keep the bricks and sand saturated with water.

- Roof over to prevent direct exposure to sun.

- Use plastic crates for storage; avoid bamboo baskets, wooden/fiber board/boxes, gunny bags etc.

- Prevent water drops coming in contact with stored material.

- Keep the chamber clean and disinfect the chamber periodically with permitted insecticide/ fungicide/ chemical, to protect from fungus, insect/ pests, reptiles etc.

Chapter 5

Vegetable and Fruit Diseases

The vegetable and fruit crops can be affected by a large number of diseases such as black dot, beet vascular necrosis, alternaria solani, tobamovirus, didymella pinodes, ralstonia solanacearum and phytophthora capsici. All these vegetables and fruits diseases have been carefully analyzed in this chapter.

Black Dot

Black dot is a common disease of potato. It is most often observed on tubers but it can affect all parts of the plant. The disease has probably been underestimated in the recent past as the symptoms are similar to more common potato diseases. On potato foliage symptoms are nearly indistinguishable from early blight and on tubers it produces blemishes that are easily mistaken for silver scurf. Although not as serious as other more common potato diseases such as black scurf, silver scurf, or common scab, it can be more devastating as it affects all parts of the plant. Above ground it can infect the vascular system causing wilt, and below ground it can cause severe rotting of roots, shoots and stolons, leading to early plant decline, discolored tubers and reduced yields.

Symptoms

Figure: Black spots on the surface of a black dot infected tuber. These are microsclerotia of the black dot pathogen *Colletotrichum coccodes*.

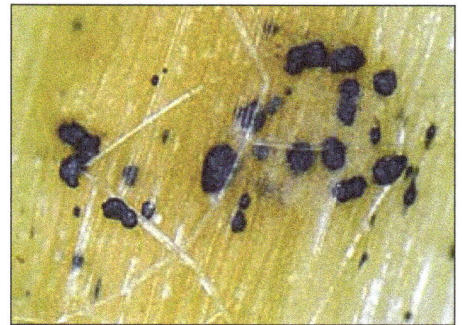

Figure: Black dot mircosclerotia on the surface of a potato root, and root hair.

Black dot is named after the abundant black dots that form on tubers. The black dots are microsclerotia that are often just visible to the naked eye. They are not restricted to tubers and can also be found on stolons, roots and stems both above and below ground. Foliar symptoms are not often observed in Michigan, but this may be due to the fact that they bear a resemblance to the small brown to black flecks characteristic of early blight. Stem lesions are more common in Michigan than foliar symptoms, especially towards the end of the growing season. Stem lesions tend to form around the base of leaf petioles. As with leaf lesions they initially start out as small brown flecks. These gradually coalesce forming lesions which may girdle the stem. As lesions mature they develop circular to irregularly shaped, white to straw-colored centers with wide margins that vary in color from brown to black. Microsclerotia form in the center of the lesions and are often clearly visible against the pale background. As infected tissue senesces, microsclerotia may become abundant covering the entire surface of the stem. Microsclerotia often appear at the base of the plant up to several inches above soil level late in the season and after vine kill.

Lesions on below ground stems and stolons may resemble Rhizoctonia lesions but are darker in appearance. Infection of root cortical tissue causes sloughing of the periderm and may result in severe rotting and early plant death. Microsclerotia and mycelia may also be abundant both internally and externally on roots and stolons.

Tuber symptoms appear as a brownish to gray discoloration over a large portion of the tuber or as circular to irregularly shaped areas. Black dot may develop a silvery sheen during storage, which can be confused with silver scurf. However, black dot tends to show much more irregularly shaped patches with less well-defined margins than silver scurf. Inspection with a hand lens (10x) will quickly differentiate the regularly spaced black dots from the bunched threads of silver scurf.

Disease Cycle

Black dot is caused by the fungus Colletotrichum coccodes. The pathogen has a wide host range, occurring on other plants in the Solanaceae family including eggplant, pepper, tomato and weeds such as hairy nightshade (Solanum sarrachoides). Colletotrichum coccodes readily produces microsclerotia on senescing plant tissue and the surface of tubers. These structures allow the pathogen to survive for long periods of time in the soil. In Michigan, the pathogen survives between growing seasons as sclerotia in infested plant debris and soil, on infected potato tubers and in overwintering debris of susceptible solanaceous crops and weeds. In the spring, sclerotia develop into acervuli. These are fruiting bodies which produce masses of spores. Spores serve as the primary inocula to initiate disease. Initial inoculum is readily moved within and between fields, as the spores are easily carried by air currents, windblown soil particles, splashing rain and irrigation water. Poor soil drainage and low plant fertility are thought to increase disease incidence and severity.

Figure: Tuber symptoms appear as a brownish gray discoloration over a large portion of the tuber.

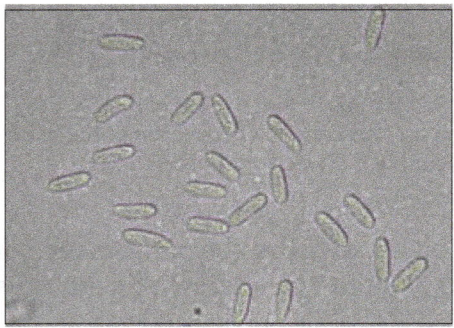

Figure: Spores of the black dot pathogen *Colletotrichum coccodes.*

Spores of C. coccodes are produced on potato plants and plant debris between 45° and 95° F. However, in greenhouse studies limited infection occurred below 59° F. As with most Colletotrichum species, C. coccodes favors temperatures above 68° F, and free moisture from rain, irrigation, fog or dew is required for spore germination and infection of plant tissues. Spores landing on susceptible plant tissue germinate and may penetrate tissues directly through the epidermis and or through wounds such as those caused by mechanical injury or insect feeding. On tomato leaves, C. coccodes has been reported to colonize lesions caused by Alternaria solani (early blight) and flea beetles.

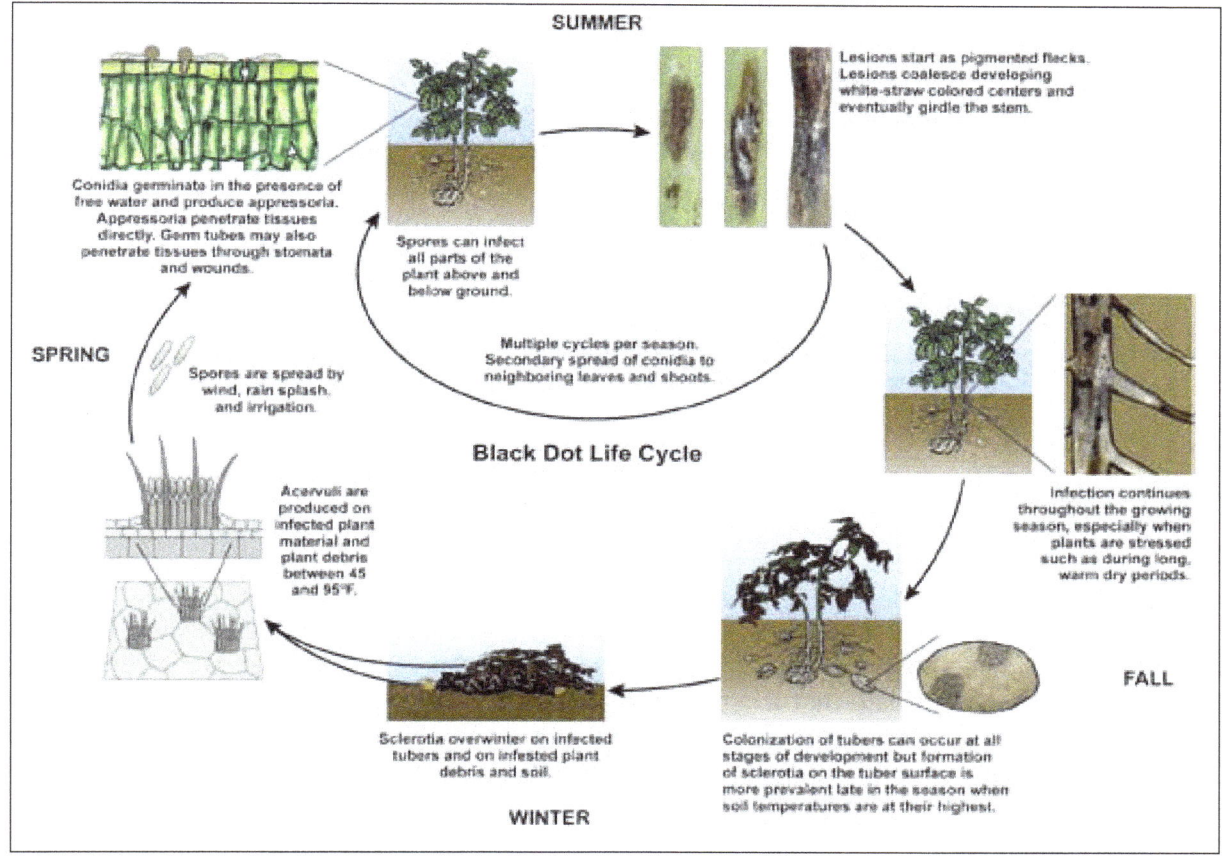

Figure: The disease cycle of the black dot pathogen, Colletotrichum coccodes.

Infection of below-ground plant parts continues throughout the growing season, especially when plants are under stress, such as during long, warm dry periods. Production of microsclerotia is greatest at high temperatures and increasing disease incidence has been associated with increasing soil temperatures. Colonization of tubers can occur at all stages of development, but formation of sclerotia on the tuber surface is more prevalent late in the season when soil temperatures are declining from the peak in high summer.

Monitoring and Control

Effective management of black dot requires implementation of an integrated disease management approach. The disease is controlled primarily through the use of cultural practices and fungicides.

Cultural Control

One of the most important approaches to black dot control is to reduce the amount of inoculum in soil through the use of cultural practices such as crop rotation, removal of crop debris, volunteer and cull potatoes from the field, and eradication of weed hosts. Since black dot microsclerotia can persist in the field for up to 2 years, a 3 to 4 year rotation with non-host crops (e.g. small grains, soy bean or corn) is often recommended to reduce the amount of inoculum in the soil. The following cultural practices are also suggested to prevent and reduce the incidence of black dot.

1. Use certified disease free seed.

2. Treat cut seed with a seed treatment (e.g. Maxim MZ). Although seed treatments may provide limited control of black dot, they improve plant stand and crop vigor, reducing plant stress which increases susceptibility to black dot.

3. Avoid planting in poorly draining soil if possible.

4. Use good crop production practices, such as timely irrigation and adequate fertilization to reduce crop stress.

5. Use tillage practices such as fall plowing that bury plant refuse and encourage decomposition.

6. Harvest tubers as soon as possible after vine kill.

7. Control temperature and humidity in storage. High temperatures and condensation on the tuber surface promotes disease.

Host Resistance

Currently there are no known commercial varieties with resistance to black dot. However, research has shown that in general late maturing varieties tend to be more vulnerable to yield reductions than early maturing varieties. This may be due to the fact that black dot is a late season disease, and leaving tubers in the ground longer exposes them to more disease pressure.

Chemical Control

In furrow applications of azoxystrobin have been reported to reduce or suppress symptoms of black dot on the stem although under conditions conducive for development of black dot, variable results have been reported. Fludioxonil, pyraclostrobin and PCNB have also been reported to suppress black dot.

Beet Vascular Necrosis

Beet vascular necrosis is a disease caused by a bacterium, Erwinia carotovora subsp. betavasculorum, present in many native and cultivated soils.

This pathogen can survive in some weedy hosts. Portions of the bacterium that causes potato blackleg can be pathogenic to sugar beets. Plant wounding, excessive nitrogen or moisture, and warm temperatures (optimum is 79° F to 82° F) favor disease development. The disease occasionally is severe in Idaho.

Symptoms:

Black streaks may be found on petioles, and crowns may be blackened or produce froth. Vascular bundles are brown, and adjacent tissue turns pink when cut and exposed to air. Rot can become extensive soft or dry rot.

Management:

Since the bacteria cannot survive in seeds, the best way to prevent the disease is to ensure that vegetatively propagated plant material are clean of infection, such that the bacterium does not enter the soil. However, if the bacteria is already present, there are some methods that can be used to lessen the infection.

Cultural Practices

Because the bacteria readily enter the plant through wounds, management practices that decrease injury to the plants are important to control the spread of the disease. Cultivation is not recommended, as the machinery can become contaminated and physically spread the bacteria around the soil. Accidental leaf tearing or root scarring can also occur depending on the size of the crop, allowing the bacteria to enter more individual plants. If hilling the beets, great care must be taken to avoid getting soil into the crown, because the pathogen is soil-borne and this could expose the plant to more bacteria, thus increasing the risk of infection.

While most bacteria are motile and can swim, they cannot move very far due to their small size. However, they can be carried along by water, and a significant movement of *Pectobacterium* can be attributed to being carried downstream from irrigation and rainwater. To control the spread of the disease, limiting irrigation is another strategy. The bacteria also flourishes in wet conditions, so limiting excess water can control both the spread and severity of the disease.

Increased in-row spacing also causes more severe disease. In an infected field, yield decreased linearly when spacing was greater than 15 cm (6 in), so a spacing of 6 inches or less is recommended.

The bacteria can also utilize nitrogen fertilizer to accelerate their growth, thus limiting or eliminating the amount of nitrogen fertilizer applied will lessen the disease severity. For example, when fertilizer was applied to an infected field the infection rate per root increased from 11% (with no added nitrogen) to 36% (with 336 kg nitrogen/hectare), and sugar yields decreased.

Resistance

The bacteria can survive in the rhizosphere of other crops such as tomato, carrots, sweet potato, radish, and squash as well as weed plants like lupin and pigweed, so it is very hard to get rid of it completely. When it is known that the bacterium is present in the soil, planting resistant varieties can be the best defense against the disease. Many available beet cultivars are resistant to *Pectobacterium carotovorum* subsp. *betavasculorum*, and some examples are provided in the corresponding table. A comprehensive list is maintained by the USDA on the Germplasm Resources Information Network. Even though some genes associated with root defense response have been identified, the specific mechanism of resistance is unknown, and it is currently being researched.

Cultivar	Resistance
H9	No
H10	No
C17	No
546 H3	Moderate
C13	No
E540	No
E538	No
E534	Moderate
E502	Moderate
E506	Yes
E536	Yes
C930-35	Moderate
C927-4	Moderate
C930-19	Yes
C929-62	Yes

Biological Control

Some bacteriophages, viruses that infect bacteria, have been used as effective controls of bacterial diseases in laboratory experiments. This relatively new technology is a promising control method that is currently being researched. Bacteriophages are extremely host-specific, which makes them environmentally sound as they will not destroy other, beneficial soil microorganisms. Some bacteriophages identified as effective controls of *Pectobacterium carotovorum* subsp. *betavasculorum* are the strains ΦEcc2 ΦEcc3 ΦEcc9 ΦEcc14. When mixed with a fertilizer and applied to inoculated calla lily bulbs in a greenhouse, they reduced diseased tissue by 40 to 70%. ΦEcc3 appeared to be the most effective, reducing the percent of diseased plants from 30 to 5% in one trial, to 50 to 15% in a second trial. They have also been used successfully to reduce rotting in lettuce caused by *Pectobacterium carotovorum* subsp. *carotovorum*, a different bacterial species closely related to the one that causes beet vascular necrosis.

While it is more difficult to apply bacteriophages in a field setting, it is not impossible, and laboratory and greenhouse trials are showing bacteriophages to potentially be a very effective control mechanism. However, there are a few obstacles to surmount before field trials can begin. A large problem is that they are damaged by UV light, so applying the phage mixture during the evening will help promote its viability. Also, providing the phages with susceptible non-pathogenic bacteria to replicate with can ensure there is adequate persistence until the bacteriophages can spread to the targeted bacteria. The bacteriophages are unable to kill all the bacteria, because they need a dense population of bacteria in order to effectively infect and spread, so while the phages were able to decrease the number of diseased plants by up to 35%, around 2,000 Colony Forming Units per milliliter (an estimate of living bacteria cells) were able to survive the treatment. Lastly, the use of these bacteriophages places strong selection on the host bacteria, which causes a high probability of developing resistance to the attacking bacteriophage. Thus it is recommended that multiple strains of the bacteriophage be used in each application so the bacteria do not have a chance to develop resistance to any one strain.

Importance

The disease was first identified in the western states of, California, Washington, Texas, Arizona and Idaho in the 1970s and initially led to substantial yield losses in those areas. *Erwinia caratovara* subsp *betavascularum* was not discovered in Montana until 1998. When it first appeared, beet vascular necrosis caused individual farm yield loss ranging from 5–70% in Montana's Bighorn Valley. Today, yield losses from the disease are generally infrequent and patchy as most producers plant resistant varieties. Infection rate is generally low if resistant cultivars are chosen; however, warmer and wetter conditions can lead to higher than normal instance of disease.

If infection does occur, bacterial root rots can not only cause economic losses in the field, but also can in storage and processing as well. In processing plants, rotten roots complicate slicing and the bacterially-produced slime can clog filters. This is especially problematic with late-infected beets which are generally harvested and processed along with healthy beets. The disease can also lower sugar-content which greatly reduces the quality.

Alternaria Solani

Alternaria solani causes diseases on foliage (early blight), basal stems of seedlings (collar rot), stems of adult plants (stem lesions), and fruits (fruit rot) of tomato.

Hosts and Symptoms

Alternaria solani infects stems, leaves and fruits of tomato (*Solanum lycopersicum* L.), potato (*S. tuberosum*), eggplant (*S. melongena* L.), bell pepper and hot pepper (*Capsicum* spp.), and other members of the *Solanum* family. Distinguishing symptoms of *A. solani* include leaf spot and defoliation, which are most pronounced in the lower canopy. In some cases, *A. solani* may also cause damping off.

On Tomatoes

On tomato, foliar symptoms of *A. solani* generally occur on the oldest leaves and start as small lesions that are brown to black in color. These leaf spots resemble concentric rings - a distinguishing characteristic of the pathogen - and measure up to 1.3 cm (0.51 inches) in diameter. Both the area around the leaf spot and the entire leaf may become yellow or chlorotic. Under favorable conditions (e.g., warm weather with short or abundant dews), significant defoliation of lower leaves may occur, leading to sunscald of the fruit. As the disease progresses, symptoms may migrate to the plant stem and fruit. Stem lesions are dark, slightly sunken and concentric in shape. Basal girdling and death of seedlings may occur, a symptom known as collar rot. In fruit, *A. solani* invades at the point of attachment to the stem as well as through growth cracks and wounds made by insects, infecting large areas of the fruit. Fruit spots are similar in appearance to those on leaves – brown with dark concentric circles. Mature lesions are typically covered by a black, velvety mass of fungal spores that may be visible under proper light conditions.

Stem lesion of *Alternaria solani*.

On Potatoes

In potato, primary damage by *A. solani* is attributed to premature defoliation of potato plants, which results in tuber yield reduction. Initial infection occurs on older leaves, with concentric dark brown spots developing mainly in the leaf center. The disease progresses during the period of potato vegetation, and infected leaves turn yellow and either dry out or fall off the stem. On stems, spots are gaunt with no clear contours (as compared to leaf spots). Tuber lesions are dry, dark and pressed into the tuber surface, with the underlying flesh turning dry, leathery and brown. During storage, tuber lesions may enlarge and tubers may become shriveled. Disease severity due to *A. solani* is highest when potato plants are injured, under stress or lack proper nutrition. High levels of nitrogen, moderate potassium and low phosphorus in the soil can reduce susceptibility of infection by the pathogen.

Disease Cycle

Alternaria solani is a deuteromycete with a polycyclic life cycle. *Alternaria solani* reproduces asexually by means of conidia. *A.solani* is generally considered to be a necrotrophic pathogen, i.e. it kills the host tissue using cell wall degrading enzymes and toxins and feeds on the dead plant cell material.

The life cycle starts with the fungus overwintering in crop residues or wild members of the Solanaceae family, such as black nightshade. In the spring, conidia are produced. Multicellular conidia are splashed by water or by wind onto an uninfected plant. The conidia infect the plant by entering through small wounds, stomata, or direct penetration. Infections usually start on older leaves close to the ground. The fungus takes time to grow and eventually forms a lesion. From this lesion, more conidia are created and released. These conidia infect other plants or other parts of the same plant within the same growing season. Every part of the plant can be infected and form lesions. This is especially important when fruit or tubers are infected as they can be used to spread the disease.

In general, development of the pathogen can be aggravated by an increase in inoculum from alternative hosts such as weeds or other solanaceous species. Disease severity and prevalence are highest when plants are mature.

Environment

Alternaria solani spores are universally present in fields where host plants have been grown.

Free water is required for Alternaria spores to germinate; spores will be unable to infect a perfectly dry leaf. Alternaria spores germinate within 2 hours over a wide range of temperatures but at 26.6-29.4°C (80-85° F) may only take 1/2 hour. Another 3 to 12 hours are required for the fungus to penetrate the plant depending on temperature. After penetration, lesions may form within 2–3 days or the infection can remain dormant awaiting proper conditions (15.5°C (60° F) and extended periods of wetness). Alternaria sporulates best at about 26.6°C (80° F) when abundant moisture (as provided by rain, mist, fog, dew, irrigation) is present. Infections are most prevalent on poorly nourished or otherwise stressed plants.

Management

Cultural Control

- Clear infected debris from field to reduce inoculum for the next year.

- Water plants in the morning so plants are wet for the shortest amount of time.

- Use a drip irrigation system to minimize leaf wetness which provides optimal conditions for fungal growth.

- Use mulch so spores in soil cannot splash onto leaves from the soil.

- Rotate to a non-Solanaceous crop for at least three years.

- If possible control wild population of *Solanaceae*. This will decrease the amount of inoculum to infect your plants.

- Closely monitor field, especially in warm damp weather when it grows fastest, to reduce loss of crop and spray fungicide in time.

- Plant resistant cultivars.

- Increase air circulation in rows. Damp conditions allow for optimal growth of *A. Solani* and the disease spreads more rapidly. This can be achieved by planting farther apart or by trimming leaves.

Chemical Control

There are numerous fungicides on the market for controlling early blight. Some of the fungicides on the market are (azoxystrobin), pyraclostrobin, Bacillus subtilis, chlorothalonil, copper products, hydrogen dioxide (Hydroperoxyl), mancozeb, potassium bicarbonate, and ziram. Specific spraying regiments are found on the label. Labels for these products should be read carefully before applying.

Quinone outside inhibitor (QoIs) fungicides e.g. azoxystrobin are used due to their broad-spectrum activity. However, decreased fungicide sensitivity has been observed in *A. solani*due to a F129L (Phenylalanine (F) changed to Leucine at position 129) amino acid substitution.

Economic Significance

Early blight caused by *A. solani* is the most destructive disease of tomatoes in the tropical and subtropical regions. Each 1% increase in intensity can reduce yield by 1.36%, and complete crop failure can occur when the disease is most severe. Yield losses of up to 79% have been reported in the U.S., of which 20-40% is due to seedling losses (i.e., collar rot) in the field.

A. solani is also one of the most important foliar pathogens of potato. In the U.S., yield loss estimates attributed to foliar damage, which results in decreased tuber quality and yield reduction, can reach 20-30%.In storage, *A. solani* can cause dry rot of tubers and may also reduce storage length, which both of which diminish the quantity and quality of marketable tubers.

Because *A. solani* is one of numerous tomato/potato pathogens that are typically controlled with the same products, accurately estimating both the total economic loss and the total expenditure on fungicides for control of early blight is difficult. Best estimates suggest that total annual global expenditures on fungicide control of *A. solani* is approximately $77 million: $32 million for tomatoes and $45 million for potatoes.

Tobamovirus

Tobamovirus is a genus in the family Virgaviridae. Tobacco, tomato, potato and squash are natural hosts to this virus. Currently, there are 37 species that are known to belong to this genus. Necrotic lesions on leaves are manifestation of tobamovirus infections on such plants. Tobamoviruses that affect the cucurbits, solanaceous, brassicas and malvaceous plants are four informal subgroups in this genus. These groups differ in terms of their genome sequences and range of host plants.

These viruses are thought to have codiverged with their hosts from a common ancestor. There are at least 3 distinct clades of tobamoviruses, which to some extent follow their host ranges: that is, there is one infecting solanaceous species; a second infecting cucurbits and legumes and a third infecting the crucifers.

Genome

The RNA genome encodes at least four polypeptides: these are the non-structural protein and the read-through product which are involved in virus replication (RNA-dependent RNA polymerase, RdRp); the movement protein (MP) which is necessary for the virus to move between cells and the coat protein (CP). The read-through portion of the RdRp may be expressed as a separate protein in TMV. The virus is able to replicate without the movement or coat proteins, but the other two are essential. The non-structural protein has domains suggesting it is involved in RNA capping and the read-through product has a motif for an RNA polymerase. The movement proteins are made very early in the infection cycle and localized to the plasmodesmata, they are probably involved in host specificity as they are believed to interact with some host cell factors.

Structure

Tobamoviruses are non-enveloped, with helical rod geometries, and helical symmetry. The diameter is around 18 nm, with a length of 300-310 nm. Genomes are linear and non-segmented, around 6.3-6.5kb in length.

Genus	Structure	Symmetry	Capsid	Genomic arrangement	Genomic segmentation
Tobamovirus	Rod-shaped	Helical	Non-enveloped	Linear	Non-Segmented

Life Cycle

Viral replication is cytoplasmic. Entry into the host cell is achieved by penetration into the host cell. Replication follows the positive stranded RNA virus replication model. Positive stranded RNA virus transcription is the method of transcription. Translation takes place by suppression of termination. The virus exits the host cell by monopartite non-tubule guided viral movement. Plants serve as the natural host. Transmission routes are mechanical.

Genus	Host details	Tissue tropism	Entry details	Release details	Replication site	Assembly site	Transmission
Tobamovirus	Plants	None	Unknown	Viral movement	Cytoplasm	Cytoplasm	Mechanical

Routes of Infection

The infection is localized to begin with but if the virus remains unchallenged it will spread via the vascular system into a systemic infection. The exact mechanism the virus uses to move throughout the plant is unknown but the interaction of pectin methylesterase, a cellular enzyme important for cell wall metabolism and plant development, with the movement protein has been implicated.

Leveillula Taurica

Powdery mildew is a serious fungal threat to agricultural production. The endoparasitic powdery mildew fungus *Leveillula taurica* (Lév.) G. Arnaud (anamorph: *Oidiopsis taurica* (Lév.) E. S. Salmon) is an important pathogen of pepper, tomato, eggplant, onion, cotton, and other crops, and it has also been recorded on many wild plant species. This pathogen represents a challenge from many perspectives. First, the early stages of infection are difficult to diagnose; thus, the disease can rapidly spread in both field and greenhouse crop production. Second, the species delimitation in the genus *Leveillula* is problematic and the binomial "*L. taurica*" clearly refers to a species complex that includes several biological species. Moreover, the exact host ranges of the different *L. taurica* lineages recognized by phylogenetic studies are still not known.

Most powdery mildew species are epiparasitic because all their structures except haustoria are developed on the host plant surfaces. In contrast, *Leveillula* and the other genera in tribe *Phyllactinieae,* namely *Phyllactinia* and *Pleochaeta*, develop a partly endophytic mycelium and are endotrophic because their haustoria are produced in the mesophyll cells. Kunoh illustrated that, in *L. taurica*, the germinated conidia are attached to the leaf surface by "adhesion bodies" that differ from appressoria produced by epiphytic species because they do not initiate infection hyphae that penetrate the epidermal cells. Instead, infection hyphae of *L. taurica* enter the host plant through stomata and develop an intercellular mycelium in the spongy and palisade parenchyma tissue with haustoria in some of its cells. Later, conidiophores emerge through stomata, mainly on the abaxial leaf surface, producing primary and numerous secondary conidia that differ in their morphology. At that stage, hyphae are also produced mainly on the abaxial leaf surfaces. On the adaxial surface of infected leaves, chlorotic spots are usually visible, indicating the development of mildew colonies underneath the spots.

Hosts and Symptoms

L. taurica is the pathogen responsible for powdery mildew on onions, but it can also infect peppers, tomatoes, eggplant, cotton, and garlic. While *L. taurica* can infect many different plants it is actually very host specific. Different races of *L. taurica* can only infect certain crops, and even specific cultivars within the same crop. An accurate way to describe its host specificity is that this disease is, "a composite species consisting of many host-specific races." Symptoms of Onion Powdery Mildew (OPM) are usually seen as circular or oblong lesions that are 5 to 20 mm and have a chlorotic or necrotic appearance. The lesions appear on older leaves before the bulb of the onion begins to form, but also can occur on the younger leaves towards the end of the season. As the disease progresses signs of OPM can also be seen. On the lesions white mycelium can be found with conidiophores bearing either lanceolate or rounded condia.

Disease Cycle

The polycyclic disease cycle of *L. taurica* is similar to that of other powdery mildew species. It overwinters (as chasmothecia) in crop residues above the soil surface. Under favorable climatic conditions, the chasmothecia open and release ascospores, which are wind-dispersed. The ascospores enter the host through its stomata, germinate, and colonize the host's tissues with its mycelia. The pathogen then begins to produce its asexual conidia, either singly or on branched conidiophores. The conidia exit through the host's stomata and serve as a secondary inoculum to spread disease after initial infection. In the fall, the pathogen undergoes sexual reproduction and again produces chasmothecia, its dormant, overwintering structure.

Environment

The genus *Leveillula* is distributed in warm, arid areas of Africa, Asia, South America, southern Europe, and the western parts of North America. Species within the genus are adapted to xerophytic conditions, exemplified by the ability of their conidia to germinate rapidly and at any relative humidity. *L. taurica* is primarily a disease of allium species—it has been documented on onions and garlic in Israel and southeastern Europe—but can also infect other species, including cucumbers, peppers, eggplants and tomatoes. It was first reported in the western United States in 1985, infecting onions in the state of California. It has since appeared in Idaho, the state of Washington, and Utah.

Management

OPM tends to appear near the end of the growing season. The best way to control *L. taurica* is to remove all crop residue from the previous onion crop before subsequent planting. Two methods to accomplish this include deep tillage, and rotating to a non-host crop the year following an onion crop. Controlling volunteer onion sprouting (or the emergence of the previous year's onion plants) will also assist in prevention of the pathogen from carrying-over from one year to the next.

Irrigation practices can also be used to limit the development of OPM. Moisture stress has been noted to increase the susceptibility of host species to *L. taurica*. Onions with adequate moisture will be more resistant to the pathogen, and onion crops with overhead irrigation rarely see powdery mildew development in the field.

The fungicide Cabrio (produced by BASF Chemical) is labeled for the control of *L. taurica* on onions, but the disease rarely progresses enough to justify the use of a fungicide. Considerations of economic benefit should be made before the fungicide is applied, and all labeling directions followed.

Resistant varieties have been found in some studies, Jahn et al. found powdery mildew resistance to be extremely beneficial in cucurbits, reducing the need for fungicide, and reducing agricultural losses due to powdery mildew pathogens. Although a truly resistant variety has not been found for onion plants, some onion genotypes with glossy leaves had selective susceptibility to *L. taurica*. Onions with the glossiest leaves were found to be most susceptible, while onions with less glossy leaves showed limited susceptibility. However, the study was unable to come to a conclusion on which variety was best suited for *L. taurica* resistance.

Importance

The economic importance of OPM is limited, as the disease is sporadic, and it rarely progresses enough to make fungicide treatment necessary. Because of the limited importance of OPM, data on incidence rates are not well documented. Simple cultural controls, as mentioned above, are usually effective in controlling losses associated with the disease. The disease geography within the United States is limited to Idaho, Utah, California, and the Pacific Northwest. Findings have also occurred in Israel, Italy, Iran, Sudan, Brazil, and Southeastern Europe.

Didymella Pinodes

Didymella pinodes (syn. *Mycosphaerella pinodes*) is a hemibiotrophic fungal plant pathogen and the causal agent of ascochyta blight on pea. It is infective on several species such as *Lathyrus sativus, Lupinus albus, Medicago spp., Trifolium spp., Vicia sativa,* and *Vicia articulata,* and is thus defined as broadrange pathogen.

Symptoms

Symptoms include lesions on leaves, stem and pods of plants. The disease is difficult to distinguish from blight caused by *Ascochyta pisi,* though *D. pinodes* is the more aggressive of the two pathogens.

Epidemiology

The disease cycle starts with dissemination of ascospores after which germination pycnidia rapidly develop. Pycnidiaspores quickly disperse by rain splashes are responsible for reinfection over short distances. Consequently, production of pseudothecia is initiated on senescent tissues. After rainfall, ascospores are released from the pseudothecia and disperse by wind over long distances.

Disease Management

Useful levels of resistance remain to be determined and the application of fungicidal sprays was reported to be uneconomical. Furthermore, reports showed that insensitivity arises against chemicals such as strobilurons after continuous application. Thus, cultural management is the preliminar option to control the disease progress by minimizing inoculum carry over as well as survival of inoculum on crop residues and in soil, and avoiding initial infection from arial inoculum. Furthermore, burying of infected residues declines pathogen survival, however, crop rotation and tillage regimes have little influence on disease severity. Delayed sowing by 3–4 weeks reduces ascochyta blight severity by more than 50%, however, such measures are not feasible at higher latitudes, because of a shorter growing season.

Host Resistance

So far, only incomplete resistance is available in the pea germplasm and quantitative differences are highly influenced by environmental conditions, plant age and physiological characteristics of plants. Tall cultivars with more erect growth suffer lower *D. pinodes* infection. Susceptibility increases with earliness and along with maturity of plants.

Besides morphological traits, a proteomic and metabolomic study pinpointed molecular markers contributing to resistance. Disease severity of leaves was also reported to be lower when pea plants are associated with rhizobial bacteria that presumalby provoke so called induced sysmteic resistance.

Ralstonia Solanacearum

Ralstonia solanacearum (Smith) (formerly called *Pseudomonas solanacearum*), is a soilborne bacterial pathogen that is a major limiting factor in the production of many crop plants around the world. This organism is the causal agent of brown rot of potato, bacterial wilt or southern wilt of tomato, tobacco, eggplant, and some ornamentals, and Moko disease of banana.

Symptoms

Above-ground symptoms include wilting of 1-2 leaves on young plants during the heat of the day. Such plants tend to recover at night. On large-leafed plants, only the tissue on one side of the mid-vein may wilt. This is very characteristic for plants such as Nicotiana. Affected leaves turn yellow and remain wilted after a time. The area between leaf veins dies and browns. Usually the main stem of the affected plants remains upright even though all the leaves may wilt and die.

Internal symptoms include light tan to yellow-brown discoloration of the vascular tissue. Long sections of infected stems reveals dark brown to black streaking in the vascular tissue as the disease progresses. As invasion proceeds, the pith and cortex of the stem become dark brown.

Symptoms in geraniums are very similar to those caused by the bacterial blight pathogen, *Xanthomonas campestris* pv. pelargonii (*Xcp*) However, while *Xcp* can cause leaf spotting, *Ralstonia* does not.

Signs of the Pathogen

Slimy, sticky ooze forms tan-white to brownish beads where the vascular tissue is cut. When an infected stem is cut across and the cut ends held together for a few seconds, a thin thread of ooze can be seen as the cut ends are slowly separated. If one of the cut ends is suspended in a clear container clean water, bacterial ooze will form a thread in the water.

Management

Growing and propagating from pathogen-free plant material is the main way to avoid problems with *Ralstonia*, regardless of the race and biovar involved. Propagators must use pathogen-free potting soil or other media, establish stock plants that are tested and known to be free of the bacteria, train workers handling the stock plants in methods and procedures that prevent the pathogen from contaminating the potting soil or coming in contact with the stock plants, and then maintaining this system throughout the propagation phase of crop production.

There are no chemicals or biological agents that adequately control these bacteria. Infected plants must be discarded as soon as possible.

As is the case with all pathogens carried on vegetatively propagated crops, the purchaser of cuttings or pre-finished plants must isolate all new, incoming plants as if the health of the plants were unknown, even if the plants have been certified as healthy. New plants must not be commingled or dispersed among other plants in the greenhouse from other sources. This procedure is crucial because by keeping plants originating from one source together allows you to observe those plants as a group, detect any abnormalities within that group, and treat or discard those plants as a group without affecting or damaging plants from other sources. Keeping them together as a group in a defined area of the greenhouse also limits the area that may need to be quarantined, sanitized, or isolated should a pathogen requiring a 'stop sale' (such as *Ralstonia solanacearum*) be found.

Meloidogyne Enterolobii

Meloidogyne enterolobii was originally described from a population collected from the pacara ear-pod tree (*Enterolobium contortisiliquum* (Vell.) Morong) in China in 1983. In 2001 it was reported for the first time in the continental USA in Florida. *M. enterolobii* is now considered as one of the most important root-knot nematode species because of its ability of reproducing on root-knot nematode-resistant (Mi-1 gene carrying genotypes) bell pepper and other economically important crops.

Management

The most efficient control method is preplant soil fumigation with methyl bromide (Mbr). That can reduce the *M. incognita* reproduction by almost 100%. However, the soil fumigant methyl bromide has been phased out in 2005 because of its negative effects on the ozone layer. A 1995 economic study declared that banning methyl bromide without an alternative method of controlling nematodes would cost the nation's bell pepper industry $127 million in losses.

Some Mbr alternatives have been tested, such as Metham sodium plus chloropicrin (Mna+Pic) and 1,3-Dichloropropene (1,3-D) plus Pic. Mna+Pic provided equal or better *Meloidogyne* control than methyl bromide plus pic, for sting nematode, they are equal to MBR plus pic. Other alternative such as Multiguard, which is a formulation of furfural, a compound derived from sugarcane waste, which has been reported to have both nematicidal and antifungal properties.

Nematode-resistant bell pepper cultivar is another method to control nematode population. Two bell pepper cultivars, *Carolina Wonder* and *Charleston Belle*, have been widely planted in the United States. However, while these varieties offer resistance to *M. incognita*, they are susceptible to *M. enterolobii*.

Crop rotation can be used to control *M. enterolobii*. The root-knot resistant bell peppers are not suggested to be planted in the field all over the seasons because that will select more *M. enterolobii*, which will survive and become a big population. Meanwhile, less severe yield loss of susceptible bell peppers has been observed when growing them after resistant bell pepper.

Alternaria Dauci

A. dauci causes leaf spot and blight in carots. It is present in all carrot production areas of the world, and is capable of rapidly causing severe foliar epidemics.

Hosts and Symptoms

Alternaria Leaf Blight is a foliar disease of carrots caused by the fungus *Alternaria dauci*. *Alternaria dauci* is included in the porri species group of *Alternaria*, which is classified for having large conidium and a long, slender filiform beak. Because many of the members of this group have similar morphology, *Alternaria dauci* has also been classified as formae specialis of carrots, or

A. porri f. sp. dauci. It has been well established that the host range of this disease is on cultivated and wild carrot, but it has also been claimed that *Alternaria dauci* has the ability to infect wild parsnip, celery, and parsley. A study in 2011 by Boedo et al. evaluated the host range of *Alternaria dauci* in a controlled environment and concluded that several non-carrot species could constitute alternate hosts, such as *Ridolfia segetum* (corn parsley) and *Caucalis tenet* (hedge parsley). Despite their findings, reports of *A. dauci* colonization on non-carrot hosts continues to be debated because the use of Koch's Postulates on recovered isolates of *A. dauci* is challenging and is rarely reported; in addition, few reports are often made of such infections in field settings.

Symptoms of *A. dauci* appear first as greenish-brown, then water-soaked, and finally necrotic lesions 8–10 days following an infection event. These lesions will appear on carrot leaflets and petioles, and have a characteristic chlorotic, yellow halo. The lesions can be irregularly shaped, and will often appear on older leaves first. Older leaves are the most susceptible to infection; when approximately 40% of the leaf surface area has become infected by *Alternaria dauci*, the leaf will completely yellow, collapse, and die. It is during extended conditions of warm, moist weather that lesions can coalesce and cause entire tops of carrot plants to die off, a phenomenon that is sometimes mistaken for frost damage. The symptoms of this disease are also commonly confused with Cercospora Leaf Blight of carrots as well as bacterial blight, and microscopic analysis is frequently needed to accurately diagnose the pathogen. *A. dauci* produces characteristically dark to olive-brown hyphae and elongated conidiophores, with conidia typically borne singly. Petiole infection can also occur without any lesion development on leaflets, and *A. dauci* can additionally result in damping-off of seedlings, seed stalk blight, and inflorescence infections. These symptoms can significantly reduce yield due to lost photosynthetic activity, prevention of mechanical harvest, and infection of commercial carrot seeds.

Disease Cycle

Alternaria dauci lesion.

Sexual reproduction of *Alternaria dauci* is not known to occur, and the disease is most active during spring, summer, and autumn cropping cycles. The disease cycle begins when fungus overwinters on or in host seed and in soil-borne debris from carrot. *A. dauci* may also be spread into fields via contaminated carrot seeds during cultivation. Once introduced, the pathogen can persist in carrot debris or contaminated seeds in the soil for up to two years. Seedling infection near the hypocotyl-root junction (just below the soil line) then occurs in the early spring following over-wintering of *Alternaria dauci* mycelium or conidia. This infected region will become necrotic and lead to the production of more asexual conidia on conidiophores, which will serve as secondary inoculum. Wind and rain cause conidia to disperse to neighboring host species, and multiple ger-mination tubes will be produced from each conidium that successfully colonizes a new host. As penetration occurs, *Alternaria dauci* will produce a chemical known as phytotoxin zinniol, which degrades cell membranes and chloroplasts, ultimately leading to the chlorotic symptoms charac-teristic of the disease. These germination tubes will pierce host cell walls to initiate infection, or if wounds are present the pathogen may enter in that manner. The process of germination, penetra-tion, and symptom development generally occurs in a timespan of 8 to 16 days, but the presence of wounds shortens the amount of time needed to carry out the process.

Following these events, conidia are repeatedly produced from leaf and stem lesions throughout the summer months, allowing the pathogen to be dispersed to its surrounding environment. Inflo-rescence that is infected by *A. dauci* early in the summer will produce nonviable seeds, but plants infected later in the summer or early fall may still carry viable seeds; this fungus remains in the pericarp and does not penetrate the embryo or endosperm (non-systemic). Following harvest in the fall, *Alternaria dauci* will persist in remaining carrot debris in the soil or be concentrated in infected seedlings, and the disease cycle will be repeated.

Environment

Production and transmission of *Alternaria dauci* is heightened during moderate to warm tem-peratures and extended periods of leaf wetness due to rainfall, dew, or sprinkler irrigation. Infec-tion can occur between temperatures of 57 - 95 degrees Fahrenheit, with 82 degrees Fahrenheit being optimal. Mycelium and spores are spread through splashing rain, tools for cultivation or contaminated soils. *Alternaria* diseases, in general, tend to infect older, senescing tissues, and on plants developing under stress. A study conducted by Vital et al. in 1999 assessed the influence of the rate of soil fertilization on the severity of Alternaria Leaf Blight in carrots and found low levels of nitrogen and potassium increased the severity of the pathogen, while high levels of the nutri-ents reduced disease severity. Though it is not well understood why this occurs, it is postulated that higher nitrogen levels may extend a plant's vigor and delay maturation, which is important because *A. dauci* is more likely to infect senescing tissue.

Management

Effective management for *Alternaria dauci* involves preventing the introduction and development of the disease. One of the best practices to avoid infection is to plant pathogen-free seed or seed treated with hot water at 50 degrees Celsius for twenty minutes. In addition to seed treated with fungicide or hot water, once harvest is complete it is imperative to turn the carrot residue under the soil. The pathogen only survives on infected plant debris, allowing this practice to hasten decom-position of the debris. Crop rotation will allow the debris enough time to decompose. Recommen-dations vary depending on location, but 2 years is the minimum allowance for rotation. Planting carrots continuously in the same field will result in increased infection. New fields should not be located near previously infected fields in order to prevent contamination through dispersal. Dispersal can occur through multiple avenues such as rain splash, farm equipment, workers, and insects.

Cultural practices can also promote reduction of *Alternaria dauci*. They include practices that will lower the duration of leaf wetness and soil moisture. Planting on raised beds with wider row spacing has been shown to reduce soil moisture, thereby limiting the spread of the disease. Symptoms tend to be more severe on carrots that are stressed or poorly fertilized. In order to avoid more severe symptoms, keep the plants free of injury, watered, and adequately fertilized. Although resistant varieties are not available, the susceptibility of the carrot differs by variety. The varieties least susceptible vary by state, and a list of varieties appropriate to a specific area can be found through the state's extension program. In the Midwest, the University of Wisconsin and the University of Michigan have bred varieties including Atlantis, Beta III, and Chancellor that exhibit resistance.

In the absence of treated seed, there are multiple chemical sprays available to treat *Alternaria dauci*. Azoxystrobin, chlorothalonil, iprodione, pyraclostrobin and bacillus are a few common fungicides to consider for foliar application. A few brand names to look for in the Midwest include RR Endura 70 WG, Rovral, and Switch. Gibberrillic acid has shown to be equally effective as the aforementioned fungicides. However, if sprayed in excess giberrillic acid can defer nutrients from the roots to foliage, resulting in undeveloped carrots. If chemically treating plants, scouting the crop is of utmost importance. Initial threshold recommendations vary depending on location, time of year, and moisture level. Different recommendations include spraying upon first evidence of symptoms or spraying once disease has reached 25% of the foliage.

Importance

Alternaria dauci is one of two leading pathogens affecting carrots around the world. Most often found in temperate climates, the disease has been found in North America, the Netherlands, the Middle East, and even parts of Southern Asia and India. Carrot leaf blight is especially damaging in that its leaf lesions not only reduce photosynthetic area, but also weaken the leaves and petioles structurally. This makes mechanical harvesting of the carrot crop less efficient, and yields are even worse when blighted leaves have been exposed to heavy frosts.

Alternaria dauci can spread rapidly if not controlled. Between February and November 2003, when the disease first spread to Turkey, 73-85% of surveyed fields were shown to be infected. Of those fields, disease rates among individual plants ranged from 65-90% total infection within the field. The highest levels of occurrence were always in moist fields with low levels of drainage.

Aster Yellows

Aster yellows is a plant disease caused by a phytoplasma bacterium, which affects over 300 species of herbaceous broad-leafed plants. Aster yellows is found over much of the world wherever air temperatures do not persist much above 32°C (90° F). As its name implies, members of the family Asteraceae are vulnerable to infection, though the disease can also affect a variety of common vegetables like, lettuce, carrot, tomato, and celery, cereals, garden plants, and wild species.

Typical symptoms include yellowing (chlorosis) of young shoots, stiff and erect bunchy growth, greenish and distorted or dwarfed flowers, and general stunting or dwarfing. The phytoplasma lives in the phloem of infected plants and is transmitted by leafhopper insects when they feed on an infected plant and then on a healthy one. No transmission occurs through leafhopper eggs or plant seed. The phytoplasma is perpetuated in overwintering weed and crop plants, in propagative parts (bulbs, corms, tubers), and in leafhoppers in mild climates. The phytoplasma is destroyed in plants and leafhoppers subjected to temperatures of 38 to 42°C (100 to 108° F) for two to three weeks; thus, aster yellows is rare or unknown in many tropical regions.

Though the disease is not lethal, control is effected chiefly by promptly removing diseased plants and all overwintering susceptible weeds. Spraying or dusting with a contact insecticide repulses the leafhopper carriers.

Symptoms of Aster Yellows

Infected petunia did not develop to normal size or color.

- Leaves are discolored pale green to yellow or white.
- In some plants, red to purple discoloration of leaves occurs.
- Leaves may be small and stunted.
- Flowers are small, malformed and often remain green or fail to develop the proper color.
- Plants infected early in the growing season may remain small and stunted.
- Many thin, weak stems grow close together forming a witches' broom.
- Tap roots of carrots are thin, small, covered in many root hairs, and often taste bitter.

Integrated Pest Management Strategies

1. Remove diseased plants: Once a plant is infected with aster yellows, it is a lost cause since the disease is incurable. Early diagnosis and prompt removal of infected plants may help reduce the spread of the disease. Although the disease itself is not fatal to the plant, its presence makes it impossible for a plant to fulfill its intended role in the garden.

2. Plant less susceptible plant species: Controlling aster yellows is difficult. As long as infected leafhoppers are around, they can infect plants. A practical way to avoid having problems with this disease is to grow plants that are not as susceptible to aster yellows. Verbena, salvia, nicotiana, geranium, cockscomb, and impatiens are among the least susceptible plants.

3. Control insects: Vegetable growers may protect susceptible crops by using the mesh fabrics that keep leafhoppers and other insects away from the plants. Some growers put strips of aluminum foil between rows because bright reflections of sunlight confuse the leafhoppers.

4. Control weeds: Remove weeds in your lawn, garden, and surrounding areas, including plantain and dandelion that may harbor the disease.

Phytophthora Capsici

Phytophthora capsici infects more than 50 plant species in more than 15 families. Among the affected plants, cucurbits and peppers are the most susceptible hosts.

Symptoms and Signs

Phytophthora blight, caused by the oomycete plant pathogen *Phytophthora capsici*, can develop on cucurbit plants at any stage of development. The pathogen can infect seedlings, vines, leaves, and fruit. The infection usually appears first in low areas of the fields where soil remains wet longer.

Damping-off: *Phytophthora capsici* causes pre- and post-emergence damping-off in cucurbits under wet and warm [20-30°C (68-86° F)] soil conditions. In seedlings, a watery rot develops on the hypocotyl at or near the soil line, resulting in plant death. Mature plants show symptoms of crown rot. Post-emergence plant death is preceded by plant wilting: a sudden, permanent wilt of the plant without a change in color of the foliage. Leaf wilting progresses from the base to the extremities of the vines. Plants often die within a few days of the first symptoms expression or after soil is saturated by excessive rain or irrigation. The stems of infected plants turn light to dark brown near the soil line and become soft and water-soaked. Infected stems collapse and die. The taproot and lateral roots of infected processing pumpkin plants usually do not exhibit symptoms. Following death of the foliage, roots may give rise to new vines if environmental conditions become less conducive for disease development. Phytophthora damping-off may result in partial to total loss of the crop.

Vine blight: Vines can be affected at any time during the growing season. Water-soaked lesions develop on vines. The lesions are dark olive and then become dark brown in a few days. Lesions girdle the stem, resulting in rapid collapse and death of foliage above the lesion.

Leaf symptoms: *Phytophthora capsici* can infect both the petioles and the leaf blades of plants. Dark brown, water-soaked lesions develop on petioles (similar to lesions on vines), resulting in rapid collapse of the petiole and leaf death. Infected leaf blades develop spots ranging from 5 mm (0.2 in.) to more than 5 cm (2 in.) in diameter. Infected areas are chlorotic at first, but within a few days they become necrotic with chlorotic to olive-green borders. Under wet and warm conditions, leaf spots expand rapidly, coalesce, and may cover the entire leaf. Under dry conditions, leaf spots cease to expand.

Fruit rot: Fruit rot can occur at any time from fruit set until harvest. Fruit rot generally starts on the site of the fruit that is in contact with the ground. However, occasionally infections will begin in other locations on the fruit where infected leaves or vines come into contact with a fruit. Also, symptoms on the upper surface of the fruit develop following rain or overhead irrigation, which can splash water containing the pathogen onto neighboring plants. Fruit rot also can develop after harvest, during transit or in storage. Fruit rot typically begins as a water-soaked lesion. Lesions expand, and become covered with white mold. The pathogen produces numerous sporangia on most infected fruit. Fruit infection progresses rapidly, resulting in complete collapse of the fruit. Phytophthora foliar blight and fruit rot may result in total loss of the crop.

Pathogen Biology

Phytophthora capsici is classified in the family Pythiaceae, order Peronosporales, and class Oomycetes. Oomycetes are not true fungi and have been placed in the kingdom Stramenopila. They are more closely related to brown algae than to true fungi. The pathogen produces asexual sporangia and biflagellate zoospores and sexual oospores. Mycelia are coenocytic (non-septate).*Phytophthora capsici* grows at 10 to 36°C (50 to 97° F), with optimal temperatures of 24 to 33°C (75-91° F). This pathogen grows rapidly on lima bean agar, and the colony diameter can reach up to 8 cm (3 in.) in 5 days. The growth patterns of colonies can vary from cottony, petaloid, rosaceous, to stellate (star-shaped).

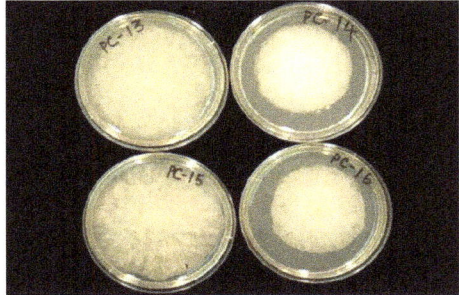

Sporangia (asexual fruiting bodies) of *P. capsici* are produced on sporangiophores (sporangia-producing hyphae) and are mostly papillate (having a small rounded protuberance). Sporangial shapes are influenced by light and other cultural conditions, and may appear as sub-spherical, ovoid, obovoid, ellipsoid, fusiform, or pyriform. The lengths and widths of sporangia can vary from 32.8 to 65.8 and 17.4 to 38.7 µm, respectively. Length/width ratios of sporangia range from 1.3:1 to 2.1:1. Sporangia have long pedicels (stalks), ranging from 35 to 138 µm. Pedicellate sporangia can be dispersed in wind driven rain. Under moist conditions, zoospores (asexual spores) are produced inside sporangia. Zoospores are single-celled and biflagellate.*Phytophthora capsici* also produces chlamydospores (thick-walled asexual spores), which may be terminal or intercalary (between cells) on the mycelium. Chlamydospores can range in diameter from 22 to 39 µm.

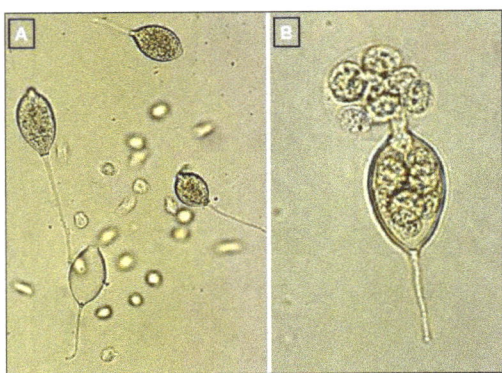

Phytophthora capsici produces sexual structures called antheridia and oogonia, and sexual spores called oospores. *Phytophthora capsici* is predominantly heterothallic with two mating types known as A1 and A2. Antheridia are amphigynous (forming a collar at the base of the oogonium after the young oogonium grows through it), with diameters of 12–21 to 12–17 µm. Oogonia are spherical or sub-spherical, with diameters ranging from 23 to 50 µm. Oospores are predominantly plerotic (filling the oogonium) with wall thicknesses ranging from 2 to 6 µm, and diameters ranging from 22 to 35 µm.

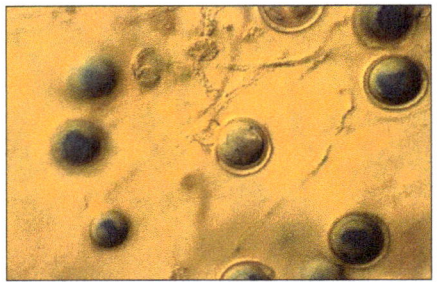
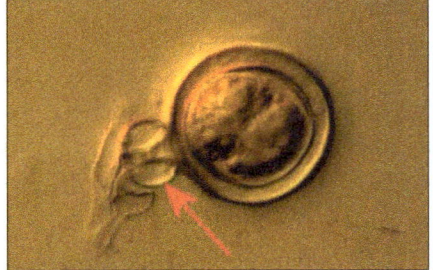

Phytophthora capsici is distinguished from other *Phytophthora* species by its sporangial morphology. Sporangia of *P. capsici* are caducous (easily separated from sporangiophores), have long pedicels, and are spherical to elongate with a tapering base.

Significant differences in virulence (degree of pathogenicity) and genetics among isolates of *P. capsici* have been reported. Several methods can be used to study the genetic variation of *P. capsici* and other fungi. Sequencing and/or restriction digest of internal transcribed spacers (ITS) regions can be used for species identification. A specific PCR primer (Pcap) has been developed that can be used with iTS primers to specifically amplify *P. capsici*. Inter-simple sequence repeats (ISSR) amplification, amplified fragment-length polymorphism (AFLP), allozyme genotyping, and restriction fragment length polymorphisms with a probe can be used to study genetic variation among populations of *P. capsici*.

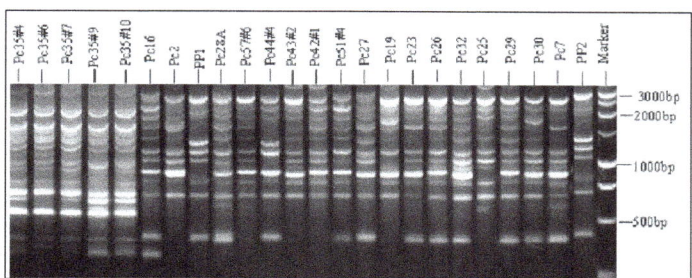

Disease Cycle and Epidemiology

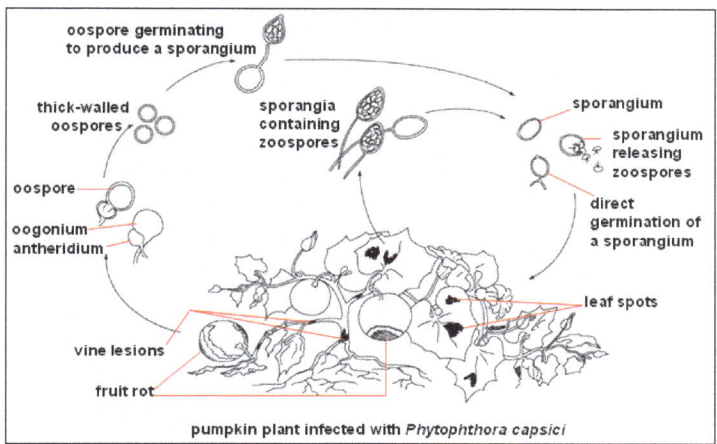

pumpkin plant infected with *Phytophthora capsici*

Phytophthora capsici is a soilborne pathogen and survives between crops as oospores in soil or mycelium in plant debris. Oospores are resistant to desiccation, cold temperatures, and other extreme environmental conditions, and can survive in the soil, in the absence of a host plant, for several years. Oospores germinate and produce sporangia and zoospores. Zoospores are released in water and dispersed by irrigation or surface water. Zoospores are able to swim for several hours and infect plant tissues. Zoospores first lose their flagella and then encyst and form a cell wall, germinate and infect plant tissues. Abundant sporangia are produced on infected tissues, particularly on affected fruit. Sporangia are dispersed by water or in wind-driven rain in the air. Sporangia may either germinate directly and infect the host plant or germinate and give rise to zoospores that are released in water and infect the plant. The pathogen grows within the host and produces sporangia on the surface of the infected tissues. If the environmental conditions are conducive, the disease develops rapidly. Although the pathogen produces chlamydospores on culture media, their role in pathogen survival and diseases epidemiology is not known.

Soil moisture conditions are important for disease development. Sporangia form when soil pores are drained, and they release zoospores when soil is saturated (soil pores are filled with water). The disease is usually associated with heavy rainfall, excessive-irrigation, or poorly drained soil. Frequent irrigation increases the incidence of the disease. Warm conditions are favorable for disease development.

Disease Management

No single method is available to provide adequate control of Phytophthora blight. Various disease control practices can be integrated to manage Phytophthora blight, including: exclusion, cultural practices, and chemical control.

Exclusion

The most effective method of control for Phytophthora blight is to prevent *P. capsici* from moving into a non-infested field.*Phytophthora capsici* spreads by soil, water, and/or plant material. It is highly recommended to thoroughly clean all farm equipment that is used in an infested field before moving it to another field. Also, avoid using water sources (i.e. ponds or reservoirs) that receive run-off water from an infested field. Water sources can be tested for the presence of the pathogen by baiting techniques.*Phytophthora capsici* is not considered a seed-borne pathogen, however, saving seed from a field where Phytophthora blight occurred should be avoided.

Cultural Practices

The following cultural practices can help to manage Phytophthora blight in cucurbit fields. Because *P. capsici* can survive in soil for several years, fields without a history of Phytophthora blight should be selected for planting. Although no cropping rotation period has been established for effective management of Phytophthora blight of cucurbits, it is recommended to select only fields that have not had a history of cucurbits, eggplant, peppers, and/or tomatoes for at least 3 years. Fields should be selected that are well isolated from fields infested with *P. capsici*. High soil moisture favors the development of Phytophthora blight, thus well-drained fields should be selected and excessive irrigation should be avoided. Also avoid planting cucurbit crops in areas of the field that have poor drainage.

Non-vining cucurbit crops (e.g. summer squash) should be planted on dome-shaped raised beds [approximately 25 cm (10 in. high)]. The field should be scouted regularly for Phytophthora symptoms, especially after major rainfalls, and particularly in low areas of the field. When symptoms are localized in a small area of the field, the infected plants should be plowed into the soil. Plants should be sprayed with effective fungicides at the first sign of the disease. Healthy fruit should be removed from the infested area as soon as possible, and they should be checked for disease development routinely. Growing cover crops and/or mulching with plant materials including straw and rye vetch can also be used to manage the dispersal of the pathogen.

Chemical Control

Fungicide seed-treatment and spray-application can prevent seedling death and reduce foliar blight and fruit rot. Seed treatment with either mefenoxam [Apron XL LS at the rate of 0.42 ml / kg (0.64 fl oz/100 lb) seed] or metalaxyl [Allegiance FL at the rate of 0.98 ml /kg (1.5 fl oz/100 lb) seed] can protect seedlings of cucurbits against *P. capsici* for up to 5 weeks after planting. Spray applications of dimethomorph [Acrobat 50WP at the rate of 448 g /ha (6.4 oz/A)] plus copper sulfate [e.g. Cuprofix Disperss 36.9F at the rate of 2.25 kg/ha (2 lb/A)], at weekly intervals, can provide effective protection against foliar blight and fruit rot caused by *P. capsici* in cucurbit fields. Combining Apron XL LS seed-treatment with spray-applications of Acrobat plus copper can minimize crop losses to Phytophthora blight in cucurbit fields. It is important to note that resistance to both mefenoxam and metalaxyl has occurred in some areas of the US, so the sensitivity of *P. capsici* populations should be tested before fungicide applications are chosen.

References

- Lunt DH. Genetic tests of ancient asexuality in root knot nematodes reveal recent hybrid origins. BMC Evol Biol. 2008;8:194–216. doi: 10.1186/1471-2148-8-194

- Sugar-beet-beta-vulgaris-bacterial-vascular-necrosis-rot-erwinia-root-rot: pnwhandbooks.org, Retrieved 21 June 2020

- David J. Hunt & Zafar A. Handoo (2009). "Taxonomy, identification, and principal species". In Roland N. Perry, Maurice Moens & James L. Starr. Root-knot Nematodes. CAB International. pp. 55–97. ISBN 978-1-84593-492-7

- Bacterial-wilt-ralstonia-solanacearum: extension.psu.edu, Retrieved 19 June 2020

- Gleason, Mark (May 2000). "Influence of Gibberellic Acid on Carrot Growth and Severity of Alternaria Leaf Blight". APS Journal. 84: 555–558. doi:10.1094/PDIS.2000.84.5.555

- Carrot-alternaria-leaf-blight, fact-sheets, vegetable: umass.edu, Retrieved 15 April 2020

- Farrar, James J.; Pryor, Barry M.; Davis, R. M. (August 2004). "Alternaria Diseases of Carrot". Plant Disease. 88: 776–784. doi:10.1094/PDIS.2004.88.8.776. Retrieved Oct 11, 2015

Permissions

All chapters in this book are published with permission under the Creative Commons Attribution Share Alike License or equivalent. Every chapter published in this book has been scrutinized by our experts. Their significance has been extensively debated. The topics covered herein carry significant information for a comprehensive understanding. They may even be implemented as practical applications or may be referred to as a beginning point for further studies.

We would like to thank the editorial team for lending their expertise to make the book truly unique. They have played a crucial role in the development of this book. Without their invaluable contributions this book wouldn't have been possible. They have made vital efforts to compile up to date information on the varied aspects of this subject to make this book a valuable addition to the collection of many professionals and students.

This book was conceptualized with the vision of imparting up-to-date and integrated information in this field. To ensure the same, a matchless editorial board was set up. Every individual on the board went through rigorous rounds of assessment to prove their worth. After which they invested a large part of their time researching and compiling the most relevant data for our readers.

The editorial board has been involved in producing this book since its inception. They have spent rigorous hours researching and exploring the diverse topics which have resulted in the successful publishing of this book. They have passed on their knowledge of decades through this book. To expedite this challenging task, the publisher supported the team at every step. A small team of assistant editors was also appointed to further simplify the editing procedure and attain best results for the readers.

Apart from the editorial board, the designing team has also invested a significant amount of their time in understanding the subject and creating the most relevant covers. They scrutinized every image to scout for the most suitable representation of the subject and create an appropriate cover for the book.

The publishing team has been an ardent support to the editorial, designing and production team. Their endless efforts to recruit the best for this project, has resulted in the accomplishment of this book. They are a veteran in the field of academics and their pool of knowledge is as vast as their experience in printing. Their expertise and guidance has proved useful at every step. Their uncompromising quality standards have made this book an exceptional effort. Their encouragement from time to time has been an inspiration for everyone.

The publisher and the editorial board hope that this book will prove to be a valuable piece of knowledge for students, practitioners and scholars across the globe.

Index

CPSIA information can be obtained
at www.ICGtesting.com
Printed in the USA
BVHW012039300822
645853BV00002B/93